Springer Theses

Recognizing Outstanding Ph.D. Research

Aims and Scope

The series "Springer Theses" brings together a selection of the very best Ph.D. theses from around the world and across the physical sciences. Nominated and endorsed by two recognized specialists, each published volume has been selected for its scientific excellence and the high impact of its contents for the pertinent field of research. For greater accessibility to non-specialists, the published versions include an extended introduction, as well as a foreword by the student's supervisor explaining the special relevance of the work for the field. As a whole, the series will provide a valuable resource both for newcomers to the research fields described, and for other scientists seeking detailed background information on special questions. Finally, it provides an accredited documentation of the valuable contributions made by today's younger generation of scientists.

Theses are accepted into the series by invited nomination only and must fulfill all of the following criteria

- They must be written in good English.
- The topic should fall within the confines of Chemistry, Physics, Earth Sciences, Engineering and related interdisciplinary fields such as Materials, Nanoscience, Chemical Engineering, Complex Systems and Biophysics.
- The work reported in the thesis must represent a significant scientific advance.
- If the thesis includes previously published material, permission to reproduce this must be gained from the respective copyright holder.
- They must have been examined and passed during the 12 months prior to nomination.
- Each thesis should include a foreword by the supervisor outlining the significance of its content.
- The theses should have a clearly defined structure including an introduction accessible to scientists not expert in that particular field.

More information about this series at http://www.springer.com/series/8790

Guoqiang Li

Bionic Functional Structures by Femtosecond Laser Micro/nanofabrication Technologies

Doctoral Thesis accepted by
the University of Science and Technology of China, Hefei, China

Springer

Author
Dr. Guoqiang Li
Department of Precision Instrument and Machinery
University of Science and Technology of China
Hefei
China

Supervisor
Prof. Jiaru Chu
University of Science and Technology of China
Hefei
China

ISSN 2190-5053 ISSN 2190-5061 (electronic)
Springer Theses
ISBN 978-981-13-0358-6 ISBN 978-981-13-0359-3 (eBook)
https://doi.org/10.1007/978-981-13-0359-3

Library of Congress Control Number: 2018940408

Printed on acid-free paper

This Springer imprint is published by the registered company Springer Nature Singapore Pte Ltd.
part of Springer Nature
The registered company address is: 152 Beach Road, #21-01/04 Gateway East, Singapore 189721, Singapore

Supervisor's Foreword

With the development of industry and society, more environmental problems such as atmospheric pollution resulting from the heavy use of paint and the exhaust emission, and water pollution from industrial discharge of organic solvents, off-shore oil accidents, and marine transportation have emerged as a critical worldwide issue, imposing severe environmental and ecological damage. The creatures in nature can skillfully and easily solve these thorny problems which have been bothering human beings. Thanks to the development of science and technology, people have gradually revealed the mysteries of nature, such as the structural colors of butterfly, self-cleaning effect of lotus leaf, and underwater superoleophobicity of fish scales. Very recently, the biomimetic preparation has attracted a wide range of interests, and inspired by nature, people also have imitated their structures by various methods, thus realizing the functions.

Based on the above background, this doctoral thesis by Guoqiang Li proposed the research of "Bionic Functional Structures by Femtosecond Laser Micro/nanofabrication Technologies." Laser processing has emerged as one-step and fast method without mask and harsh environment to prepare special surface micro-/nanostructures, even complex 3D structures overcoming the disadvantages of traditional methods. In this doctoral thesis, a variety of bionic functional structures are designed and produced by combining the advanced femtosecond laser micro-/nanofabrication technology and cutting-edge bionic technology.

As a supervisor of Dr. Guoqiang Li, I'm very pleased to be informed that he was entitled as reward-gainer of the excellent Ph.D. dissertations by the Chinese Academy of Sciences. We believe the original and novel works presented in this doctoral thesis will attract the readers of various research fields including micro/nanofabrication, biosurface, chemistry, physics, and would be of general interest to the broad readership.

Hefei, China
April 2018

Prof. Jiaru Chu

Abstract

Some living organisms have developed the obviously morphological and anatomical characteristics during the long period of stringent evolution to maintain habits and lifestyles in harmony with nature. Learning from nature, the living organisms in nature give us inspiration to design and create high-performance functional structures and devices to solve the tough problems in daily life.

In this work, the preparation of bionic functional micro/nanostructures was proposed by combining the advanced femtosecond laser micro/nanofabrication technology and cutting-edge bionic design. The details are given as follows:

(1) Based on the understanding of the relationship between the structures and the colors in nature, such as peacock feathers, butterfly wings, we prepared various micro/nanostructures and realized colorizing the material's surface without pigment. By adopting different laser pulse overlaps, several kinds of structural colors that are independent of the viewing angles, such as golden, white, gray, and black, were produced on the aluminum surfaces. Considering the laser-induced periodic surface structures (LIPSSs) as a kind of grating structures, we not only discussed the combined influence of incident light angle and the ripples orientation on the diversity of structural colors, but also proposed the possibility of producing multipatterns constituted by ripples with different orientations, and realized the diverse and exclusive display. In addition, an innovative method was proposed to produce three-dimensional effect patterns on metal surface by translating each layer in one direction.
(2) A facile approach was developed to prepare large-area functional polymer surface duplicated from the metal. The aluminum was firstly scanned by femtosecond laser beam with different pulse overlaps to form hierarchical structures. Then, the aluminum pieces acted as template masters and the hierarchical structures were subsequently transferred onto polydimethylsiloxane (PDMS) surfaces. It was found that the polymer replicas are endowed with tunable wetting properties. The reusability of metal master and easiness of soft lithography make it to be a very simple, fast, and cost-efficient way for mass production of functional polymeric surfaces.

(3) The capability to realize 2D–3D controllable metallic micro/nanostructures is of key importance for various fields such as plasmonics, electronics, bioscience, and chemistry due to unique properties such as electromagnetic field enhancement, catalysis, photoemission, and conductivity. However, most of the present techniques are limited to low-dimension (1D–2D), small-area, or single function. Here, we reported the assembly of self-organized three-dimensional (3D) porous metal micro/nanocage arrays on nickel surface by ethanol-assisted femtosecond laser irradiation. The underlying formation mechanism was investigated by a series of femtosecond laser irradiation under different exposure time. We also demonstrated the ability to control the size of micro/nanocage arrays from 0.8 to 2 μm by different laser pulse energy. This method features rapidness (~10 min), simplicity (one-step process), and ease of large-area (4 cm^2 or more) fabrication. The 3D cage-like micro/nanostructures exhibit not only improved antireflection from 80% to 7% but also enhanced hydrophobicity from 98.5° to 142° without surface modification.

(4) The preparation of superhydrophilic, superoleophilic, underwater superoleophobic surfaces is inspired by natural surfaces such as fish scales possessing hierarchical micro/nanostructures. We reported the assembly of self-organized hierarchical microcone arrays on a nickel surface by sucrose solution-assisted femtosecond laser irradiation. The processed surface is superhydrophilic (13.47° ~ 4.01°), superoleophilic (7.45° ~ 3.73°), and underwater superoleophobic (135.22° ~ 166.16°) which are comparable to those of fish scales. The wettabilities of the processed surfaces are tunable by adjusting the mass ratio of sucrose to water and pulse energy to control the height (1.62 ~ 10.34 μm) and size (2.1 ~ 2.81 μm) of the microcones. Multifunctional liquid manipulation such as microdroplet transferring, static and dynamic storage, liquid transportation, and mixing is demonstrated.

(5) A new approach based on liquid (ethanol and sucrose solution)-assisted femtosecond laser irradiation on silicon substrates was proposed for the preparation of controllable microcones and micromolars arrays. Their height can be controlled from 3.3 to 17.6 μm for microcones and 5.9 ~ 33.7 μm for micromolars by adjusting the pulse energy. The processed surfaces are superhydrophilic (25.05° ~ 2.46°), superoleophilic (7.22° ~ 0), and underwater superoleophobic (124.9° ~ 169.2°). The surfaces further demonstrate many distinct functions such as fog collecting and volatilizing, droplet storage and transportation, and liquid directional transfer.

(6) A kind of novel multifunctional ultrathin aluminum foil which consists of large-area regular micropore arrays covered with nanostructures was presented. These hierarchical micro/nanostructures show underwater superoleophobic ability and oil/water separation function. The novel foils were realized by one-step femtosecond laser irradiation, which is a simple and promising method for preparing special micro/nanostructures due to its high precision, excellent

controllability, one-step processing, and compatibility with various materials. In addition, the micropore-arrayed aluminum foil also shows robust filtering performance for particles with different sizes, exhibiting multifunctional applications.

Keywords

Bionic functions · Femtosecond laser processing · Micro/nanostructures · Structural colors · Superhydrophobic · Underwater superoleophobic

Parts of this thesis have been published in the following journal articles:

1. Guoqiang Li, Hua Fan, Feifei Ren, Chen Zhou, Zhen Zhang, Bing Xu, Sizhu Wu, Yanlei Hu, Wulin Zhu, Jiawen Li, Yishan Zeng, Xiaohong Li, Jiaru Chu and Dong Wu, 2016, Multifunctional ultrathin aluminum foil: oil/water separation and particle filtration. *Journal of Materials Chemistry A*, **4**, 18832–18840. **DOI:** 10.1039/C6TA08231A.
2. Guoqiang Li, Yang Lu, Peichao Wu, Zhen Zhang, Jiawen Li, Wulin Zhu, Yanlei Hu, Dong Wu and Jiaru Chu, 2015, Fish scale inspired design of underwater superoleophobic microcone arrays by sucrose solution assisted femtosecond laser irradiation for multifunctional liquid manipulation. *Journal of Materials Chemistry A*, **3**, 18675–18683. **DOI:** 10.1039/C5TA05265C.
3. Guoqiang Li, Jiawen Li, Chenchu Zhang, Yanlei Hu, Xiaohong Li, Jiaru Chu, Wenhao Huang and Dong Wu, 2014, Large-area one-step assembly of three-dimensional porous metal micro/nanocages by ethanol-assisted femtosecond laser irradiation for enhanced antireflection and hydrophobicity. *ACS Applied Materials & Interfaces*, **7**, 383–390. **DOI:** 10.1021/am506291f.
4. Guoqiang Li, Zhen Zhang, Peichao Wu, Sizhu Wu, Yanlei Hu, Wulin Zhu, Jiawen Li, Dong Wu, Xiaohong Li and Jiaru Chu, 2016, One-step facile fabrication of controllable microcone and micromolar silicon arrays with tunable wettability by liquid-assisted femtosecond laser irradiation. *RSC Advances*, **6**, 37463–37471. **DOI:** 10.1039/C6RA06949E.
5. Guoqiang Li, Jiawen Li, Liang Yang, Xiaohong Li, Yanlei Hu, Jiaru Chu, Wenhao Huang, 2013, Evolution of aluminum surface irradiated by femtosecond laser pulses with different pulse overlaps. *Applied Surface Science*, **276**, 203–209. **DOI:** 10.1016/j.apsusc.2013.03.067.
6. Guoqiang Li, Jiawen Li, Yanlei Hu, Chenchu Zhang, Xiaohong Li, Jiaru Chu, Wenhao, 2014, Realization of diverse displays for multiple color patterns on metal surfaces. *Applied Surface Science*, **316**, 451–455. **DOI**: 10.1016/j.apsusc.2014.08.030.
7. Guoqiang Li, Jiawen Li, Yanlei, Chenchu Zhang, Xiaohong Li, Jiaru Chu, Wenhao Huang, 2016, Femtosecond laser color marking stainless steel surface with different wavelengths. *Applied Physics A*, **118**, 1189–1196. **DOI**: 10.1007/s00339-014-8868-3.
8. Yanlei Hu, Guoqiang Li, Jianwen Cai, Chenchu Zhang, Jiawen Li, Jiaru Chu, Wenhao Huang, 2014, Facile fabrication of functional PDMS surfaces with tunable wettablity and high adhesive force via femtosecond laser textured templating. *AIP Advances*, **4**, **DOI:** 10.1063/1.4905052.

Contents

1 Introduction 1
1.1 Structural Colors 1
1.1.1 Common Structural Colors in Nature 1
1.1.2 Research Progress of Bionic Structural Colors 2
1.2 Basic Concept of Wettability 5
1.2.1 Contact Angle 5
1.2.2 Sliding Angle 6
1.3 Superhydrophobic Phenomena and Bionic Study 7
1.3.1 Superhydrophobic Phenomena in Nature 7
1.3.2 Preparation Methods of Superhydrophobic Surfaces 8
1.4 Underwater Surperoleophobic Surfaces 10
1.5 Main Research Contents 12
References 13

2 Key Technical Problems of Femtosecond Laser Bionic Surfaces 21
2.1 Femtosecond Laser Micro/Nanofabrication Equipment and Systems 21
2.2 Material and Their Parameters 22
2.3 Technological Process 23
2.4 The Influence of Laser Parameters and Environment on the Morphology of the Micro/Nanostructures 23
References 28

3 Study on Femtosecond Laser Induced Structural Colors 29
3.1 Femtosecond Laser Induced Structural Colors on Aluminum Surfaces 29
3.2 Diverse Displays for Multiple Color Patterns on Stainless Steel Surfaces 37

3.3 The Preparation of Three-Dimensional (3D) Patterns 44
3.4 Coloring Metal Surface with Different Laser Wavelengths 45
References 51

4 Facile Fabrication of Functional PDMS Surfaces via Femtosecond Laser Textured Templating 53
4.1 Optical and Morphology Properties of Aluminum Templates 53
4.2 Polymer Replicas of Metal Masters and Their Wetting Properties 56
4.3 Tunable Contact Angle and Strong Adhesion Force 58
Reference 60

5 Multifunctional Three-Dimensional Porous Metal Micro/ Nanocages by Ethanol-Assisted Femtosecond Laser Irradiation 61
5.1 Experimental Section 62
5.2 Formation of 3D Micro/Nanocage-like Structures 63
5.3 Precisely Controlling the Micro/Nanocages Size by Adjusting the Laser Pulse Energy 65
5.4 Enhanced Antireflection of 3D Porous Metal Micro/Nanocages 69
5.5 The Wetting Property of 3D Porous Metal Micro/Nanocages 70
References 75

6 Microcone Arrays by Sucrose Solution Assisted Femtosecond Laser Irradiation 77
6.1 Experimental Section 77
6.2 Preparation of Microcone Arrays 78
6.3 Precisely Controlling the Growth of the Cone by Adjusting the Mass Ratio of the Sucrose Solution and the Laser Pulse Energy 80
6.4 The Wettabilities of the Microcone Arrays 82
6.5 The Liquid Manipulation 87
6.5.1 Liquid Droplet Transferring 87
6.5.2 Liquid Storage 90
6.5.3 Liquid Transportation 91
6.5.4 Liquid Mixing 92
References 93

7 Fabrication of Controllable Microcone and Micromolar Silicon Arrays with Tunable Wettability by Liquid-Assisted Femtosecond Laser Irradiation 95
7.1 Experimental Section 95
7.2 Formation of Microcone and Micromolar Arrays in Ethanol and Sucrose Solutions 98
7.3 Precisely Controlling the Growth of the Microcones and Micromolars Arrays by the Laser Pulse Energy 99

7.4 The Wettability of the Microcones and Micromolars Arrays 101
7.4.1 The Superhydrophilicity/Superoleophilicity 101
7.4.2 Underwater Superoleophobicity and Oil Droplet Manipulation 103
7.4.3 The Anisotropic Wetting and Sliding Behavior 105
References 108

8 Micropore-Arrayed Ultrathin Aluminum Foil for Oil/Water Separation and Particle Filtration 109
8.1 The Preparation of the Micropore-Arrayed Aluminum Foils 110
8.2 The Wettability of the Micropore-Arrayed Aluminum Foils 112
8.3 The Oil/Water Separation by the Micropore-Arrayed Aluminum Foils 114
8.4 The Filtration of Particles with Different Diameters 122
References 124

9 Conclusions and Outlook 127

Chapter 1
Introduction

Experienced the long period of stringent evolution, the living organisms develop their unique structures and materials through natural selection, and thus adapt to the nature. Inspired by the living organisms, a new, interdisciplinary, and cutting-edge bionics which permeates and combines not just biology, computer science, but also nanotechnology, materials science, chemistry, physics has grown considerably in recent years [1–3]. Bionics refers to the application of biological methods, structures, functions, and systems found in nature to the study and design of engineering systems and modern technology for the purpose of solving complex problems which have troubled human beings for decades [1–3]. A crystallization of mankind's intelligence, bionics is a significant symbol of scientific and technological progress, which may find a wide range of applications in communications, mechanical engineering, biomedicine, artificial intelligence and so on.

1.1 Structural Colors

1.1.1 Common Structural Colors in Nature

An important branch of bionics is the study of why the living organisms produce colors and how to implement them manually. As far as the color principle is concerned, the colors can be divided into chemical color [4, 5] and physical color [6–8]. Chemical color usually refers to the color produced by the absorption of light through the pigment. The physical color is from the micro/nanostructures, namely the structural colors, and it can be explained as the interference [9–11], diffraction [12–14], and scattering [15–18] of the incident light.

Using modern testing methods, scientists have obtained the understanding of the formation mechanism of the structural colors. For example, E. Shevtsova [19] reported the striking and stable structural color patterns in the transparent wings of small hymenoptera and diptera, which is caused by thin film interference. Zi [20] reported the mechanism of color production in peacock feathers, and found that the cortex in differently colored barbules, which contains 2D photonic-crystal structures

G. Li, *Bionic Functional Structures by Femtosecond Laser Micro/nanofabrication Technologies*, Springer Theses, https://doi.org/10.1007/978-981-13-0359-3_1

are responsible for coloration of the peacock feathers. The simulated results indicate that the coloration strategies in peacock feathers are very ingenious and simple: controlling the lattice constant and the number of periods in the photonic-crystal structures.

1.1.2 Research Progress of Bionic Structural Colors

With the development of micro and nano-fabrication technology, the preparation of structural colors with artificial methods has appeared. At present, the methods mainly contain self-assembly [21], plate printing [22], and laser direct writing [21, 23–27].

For example, Chun [21] created a large, flexible reflector that actually provides better angle-independent color characteristics than Morpho butterflies and which can be bent and folded freely without losing its Morpho-mimetic photonic properties by using a combination of directional deposition, silica microspheres with a wide size distribution, and a PDMS encasing.

Kim [22] demonstrated the high-resolution patterning of multiple structural colors within seconds, based on successive tuning and fixing of color using a single material along with a maskless lithography system. They invented a material called 'M-Ink', the color of which is tunable by magnetically changing the periodicity of the nanostructure and fixable by photochemically immobilizing those structures in a polymer network. This simple, controllable, and scalable structural color printing scheme presented may have a significant impact on color production for general consumer goods.

Femtosecond (fs) laser micromachining has emerged as a new technique for micro/nano structure fabrication in recent years due to the ultra-short time scale, ultra-high peak power density characteristics of the femtosecond lasers, and its applicability to virtually all kinds of materials in an easy one-step process with very high resolution and accuracy. Vorobyev and Guo [23–27] demonstrated the preparing of structural colors on metal surfaces with femtosecond laser processing technique to provide a controllable modification of optical properties of metals from UV to terahertz via surface structuring on the nano-, micro-, and submillimeter-length scales (Fig. 1.1) Giving the additional advantages of laser processing such as low contamination and capability to process complicated shapes, the color metals have tremendous potential in various technological applications [23].

It is indicated that the focused laser beam can induce periodic structures with size typically in the order of the wavelength when the laser power is close to the damage threshold [28–31]. This periodic structure is called laser-induced periodic surface structures (LIPSS) [28–31]. In addition, it is indicated that the ripples (LIPSSs) orientation usually is perpendicular to the polarization of laser pulse [32–35]. Dusse [36] considered these LIPSSs as a kind of grating structures, which could play important roles in modifying the optical properties of metal surfaces in a very versatile way. By using nanostructures with controlled orientations, specific color patterns can be generated (Fig. 1.2).

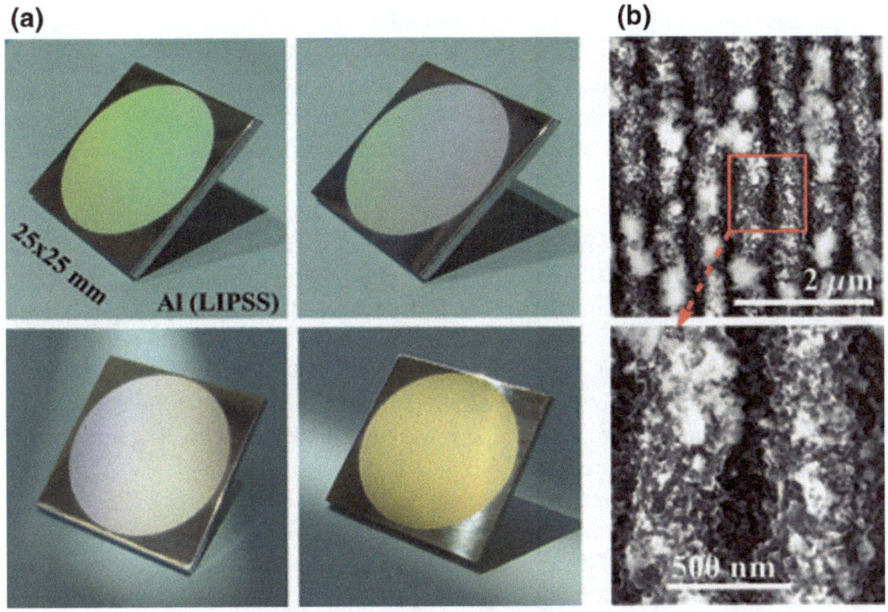

Fig. 1.1 Color aluminum due to femtosecond laser-induced LIPSSs

Fig. 1.2 Color effects by controlled nanostructures with femtosecond laser on stainless steel surface

The laser induced periodic surface structures have attracted great interest due to their wide applications in color display, anti-counterfeiting, decoration, sensing, and optical data storage etc. [36, 37]. It is found that in most cases the periods of LIPSSs are slightly smaller than laser wavelength. For color display, LIPSS are usually employed because their periods fabricated by femtosecond laser lasers with central wavelength of 800 nm can effectively diffract white light and exhibit vivid structural colors [37]. From the viewpoint of practical application, it is essential to produce deep and uniform ripples that can be extended to a large area. It is indicated that the quality of the formed ripples depends strongly on the irradiation fluence of the laser pulses and the scanning speed of the laser beam [37]. Yao [37] decorated the same or different regions of the stainless steel surface with two or more types of high quality ripples with different orientations by skillfully choose the fluence, scanning speed, and the polarization of the laser beam (Fig. 1.3). In this way, different patterns with different spatial overlapping can be selectively displayed with structural colors when illuminating with white light [37].

In addition to this, scientists have explored many other artificial methods for preparing kinds of structural colors. Moreover, the rapid development of modern manufacturing industry and the precise machining technique makes the preparation technology of structural colors more mature [37–47].

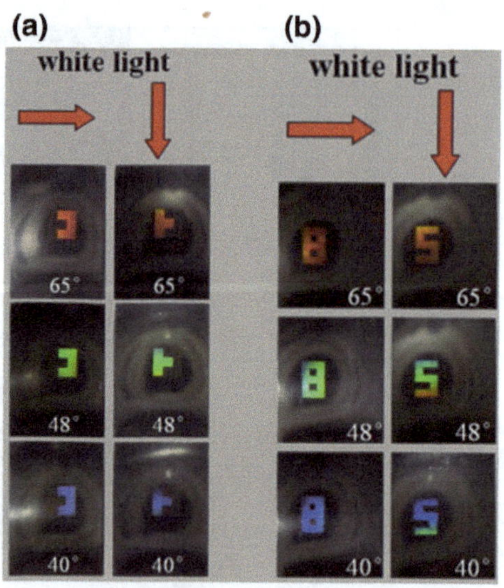

Fig. 1.3 Selective display of the two patterns by irradiating white light

1.2 Basic Concept of Wettability

1.2.1 Contact Angle

Wettability is a fundamental property of a solid surface, which refers the ability of a liquid to maintain contact with a solid surface resulting from intermolecular interactions when the two are brought together [48–50]. Contact Angle θ is an important index of the wettability of the material surface by a liquid via the Young equation [51–57]. The contact angle is related to the surface tension of the liquid and to the interfacial tension between the liquid and the solid [58–64]. When the surface tension of the liquid is less than that of the solid, the liquid will spread on the solid surface as much as possible [58–64]. On the contrary, the liquid will sit on the solid surface in a shape of sphere [58–64].

When a droplet is dropped onto the solid surface, there are several possible cases happen:

(1) The liquid spreads completely on the solid surface under the strong surface tension and the contact angle is close to 0. This kind solid surface is defined as superhydrophilic surface.
(2) When the contact angle is $10° < \theta < 90°$, the liquid partly wets the solid surface, thus it is defined as hydrophilic.
(3) When the contact angle is in the range of $90° < \theta < 150°$, the surface is hydrophobic.
(4) The superhydrophobic surfaces exhibit contact angles greater than 150°. In this case, the droplet keeps a spherical shape on the surface and can't wet it.

However, in experiment, the angle measured on the rough surface by the contact angle measurement system is apparent contact angle, and the actual contact angle is immeasurable [65, 66]. In spite of this, the apparent contact angle is still associated with the actual contact angle.

According to this situation, Wenzel [65], Cassie, and Baxter [66] proposed theoretical model for the contact angle of the rough structure on the basis of Yang's equation [67].

The Wenzel [65] model describes the homogeneous wetting regime, and is defined by the following equation for the contact angle on a rough surface:

$$\cos\theta_w = r(\gamma_{sv} - \gamma_{sl})/\gamma_{lv} = r\cos\theta_o \tag{1.1}$$

where θ_w, and θ_o are the water contact angle on the structured surface and flat surface, respectively. r is the roughness factor, which is defined as the ratio of the actual apparent surface area to geometric projected area. γ_{sv}, γ_{sl}, and γ_{lv} are the surface tensions of the solid/vapor, solid/liquid, and liquid/vapor interfaces [65]. In Wenzel state, the water droplet is in full contact with the surface, as shown in Fig. 1.4a. For the actual apparent surface area of the liquid or the solid is always greater than the geometric projected area, the value of r is greater than 1 [65]. Superhydrophobic

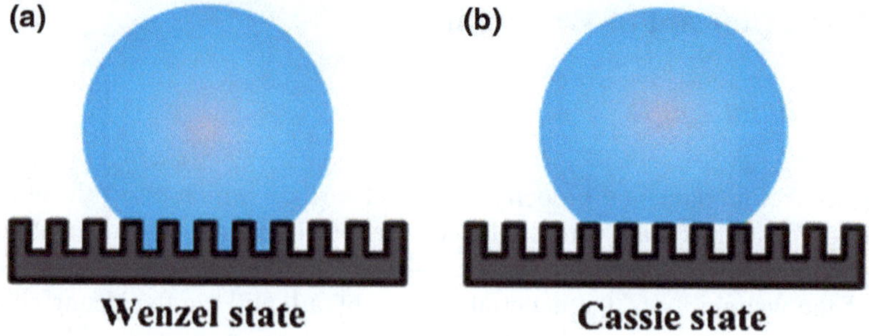

Fig. 1.4 Schematic illustration of droplet on rough substrate either in Wenzel state (**a**) or Cassie-Baxter state (**b**)

properties can be reached only if $\theta_o > 90^o$, but with high hysteresis and sliding angle due to the increase in the solid/liquid interface [65, 66]. Thus the Cassie-Baxter model was proposed as shown in Fig. 1.4b. In this model, the droplet can suspend on the composite interface made of solid and air trapped between the droplet and the surface. This model can be explained as Cassie-Baxter equation [66]:

$$\cos\theta_c = r_f f \cos\theta + f - 1 \tag{1.2}$$

where θ_c is the apparent contact angle of the solid/liquid/gas interfaces, r_f is the roughness coefficient, f is the area fraction of the liquid to the solid, and θ is the water contact angle on the flat surface. It is seen from Eq. 1.2, when $f = 1$, and $r_f = r$, Cassie-Baxter equation is equivalent to Wenzel equation.

It is found that both of Wenzel state and Cassie state involve the influence of surface roughness on hydrophobic properties. The fundamental difference between the two models is the different contact state of the droplets on the rough structures of the surface, which induces the different dynamic performances of the liquid, such as sliding angle and so on. Nevertheless, these two models reveal the relationship between the rough structure and the hydrophobicity, which provides a basis for the study of hydrophobic phenomena and its function.

1.2.2 Sliding Angle

The sliding angle, another important factor to evaluate the surface wetting property, is the critical angle where a water droplet with a certain weight begins to slide down the inclined plate [68–85]. When the droplet begins to slide down from an inclined surface, the front and back end of the droplets form the maximum contact angle (θ_{max}) and minimal contact angle (θ_{min}) under the gravitational effects. As function of maximum contact angle (θ_{max}) and minimal contact angle (θ_{min}), the sliding angle can be expressed as:

$$\sin\alpha = \gamma\frac{Rk}{mg}(\cos\theta_{\min} - \cos\theta\max) \quad (1.3)$$

where α is sliding angle, γ is surface tensions of the solid surface. m, and R are the mass, diameter of droplets, respectively. k is fitting parameter based on the experimental data, which refers to the droplet contour constants [68–85].

1.3 Superhydrophobic Phenomena and Bionic Study

1.3.1 Superhydrophobic Phenomena in Nature

The superhydrophobic surface is defined as the contact angle greater than 150° and the sliding angle less than 10°. Many living organisms have superhydrophobic surfaces which exhibit special functions [86–94], such as the lotus leaves [101–105] and the water striders [106–119] (Fig. 1.5). Water droplets from the rain and fog can roll freely on lotus leaves surfaces and remove dirt, keeping the leaves clean although many of these plants tend to grow in muddy waters [106–119]. This self-cleaning property, also known as the lotus effect, plays a critical role in the survival of the plant by keeping the surface clean and free of contamination and microoragnisms.

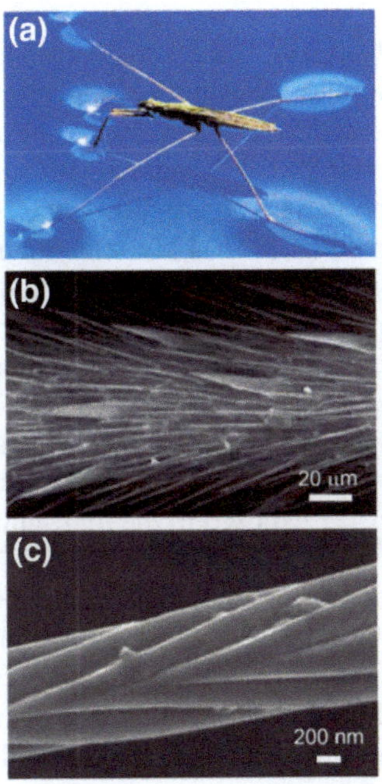

Fig. 1.5 The non-wetting leg of a water strider

Barthlott [95–97], Neinhuis [95–100], and Jiang [101–105] thought that the hydrophobicity is closely related to the surface morphology. It is indicated that the surface structures can be divided into 3D structures of varying morphologies and an underlying two dimensional wax film. 3D epicuticular wax structures occur in sizes from 0.5 to 100 μm, whereas 2D wax films range from a few molecular layers up to 0.5 μm. The structures can be illustrated that branch-like nanostructures are inlayed on the top of the micropapillae. These hierarchical surface structures endow the lotus leaves with high contact angle and low adhesive force, and demonstrate self-cleaning ability [95–105].

In addition to the lotus leaves, water striders are typical animal that possesses superhydrophobicity (Fig. 1.5a). SEM images (Fig. 1.5b and c) reveal that there are numerous oriented needle-shaped setae on the legs with diameters ranging from 3 μm down to several hundred nanometres, most length of 50 μm, and arranged at an inclined angle of about 20° from the surface of leg. In addition, many elaborate, nanoscale grooves are evident on each microseta, and these form a unique hierarchical structure [106–119]. These special hierarchical structures enable them to stand effortlessly and move quickly on torrential water. It is indicated that the maximal supporting force of a single leg is 152 dynes, which is roughly 300 times of the leg itself, indicating that its surface is strikingly water repellent [106–119].

1.3.2 Preparation Methods of Superhydrophobic Surfaces

Inspired by the lotus leaves and water spiders, people have been exploring how to imitate these surface superhydrophobic structures and studying their applications in practice. The key to the preparation of superhydrophobic surface is to obtain the micro/nano structure with high roughness. Although there are many methods of preparing superhydrophobic surface, they mainly can be summarized as template method [116–121], electrochemical method [124–130], and lithography method [131, 132]. Template method is a copying method which is mainly used to mold polymers, such as Polydimethylsiloxane (PDMS), and is more suitable for copying the structure of biological surfaces [116–121]. Sun [122] transferred the complex surface patterns of the lotus leaf onto the surface of solid-state PDMS with high fidelity by nanocasting. It is shown that both the natural and artificial lotus leaves remain superhydrophobic after being immersed in water for more than a week or being exposed to the high-humidity ambient atmosphere for several months.

Lee [123] successfully fabricated biomimetic superhydrophobic surfaces on high-density polyethylene surfaces by heat- and pressure-driven imprinting methods using patterned AAOs as replication templates. It is indicated that both nanostructures and microstructures cooperatively affect the superhydrophobicity of the polymer surface. This simple and reproducible method of producing a hierarchical superhydrophobic surface can be applied to all thermoplastic polymers [123].

The common chemical methods include electrostatic spinning [124], self-assembly [125–130], and electrochemical deposition [125]. Jiang [124] reported

Fig. 1.6 Combining layer-by-layer assembly with electrodeposition of silver aggregates for fabricating superhydrophobic surfaces

the preparation of core/shell ultrathin fibers with a novel nanowire-in-microtube structure by a multifluidic coaxial electrospinning approach. The advantage of this approach is that it introduces an extra middle fluid between the core and shell fluids of traditional coaxial electrospinning, which can work as an effective spacer to decrease the interaction of the other two fluids. Layer-by-layer assembly method [125–130] is a common self-assembly method of assembling the different layers of polyanions and polycation by static electricity. The advantage of this method is that the thickness of the generated film can be precisely controlled by adjusting the number of assembly times. This method combined with electrochemical deposition is commonly used to prepare superhydrophobic surfaces because it can improve the surface roughness by introducing the nanoparticles [125–130]. Electrochemical deposition [125] refers to the process of depositing metal alloys or metal compounds from the solution or molten salt into electrode surfaces under the action of electric field. Zhao [125] developed a novel method to fabricate a branchlike Ag micro-/nanostructure with the combination of the layer-by-layer assembly technique with electrodeposition by taking advantage of the stability and penetrability of layer-by-layer films (Fig. 1.6). After further simple surface modification, this surface with the branchlike Ag aggregate film has showed remarkable superhydrophobicity with a contact angle as high as 154° and a tilt angle lower than 3°. It is believed that this work offers new promising applications for the fabrication of nanostructured surfaces [125].

Laser and plasma etching [131, 132] is a direct method of preparing rough structures on base material surfaces. Khorasani [131] fabricated superhydrophobic and superhydrophilic surfaces on PDMS with the irradiation of a CO_2-pulsed laser, showing excellent blood compatibility because of the most and the least wettability of these surfaces. Long [132] fabricated laser-induced periodic surface structures (LIPSS), i.e. ripples, by picosecond laser nanostructuring on copper surfaces. After the modification with triethoxyoctylsilane, the specific type of ripples exhibited superhydrophobicity with a large contact angle of $153.9 \pm 3.2°$ and a low sliding angle of $11 \pm 3°$ (Fig. 1.7).

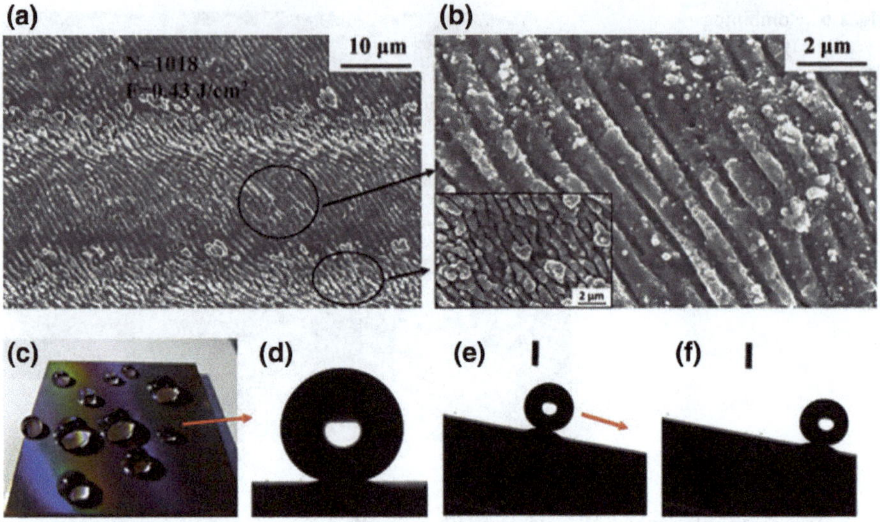

Fig. 1.7 Superhydrophobic and colorful copper surfaces fabricated by picosecond laser

With the development of femtosecond laser technology, people began to explore the preparation of superhydrophobic surface by using femtosecond laser on various material surfaces [133–137]. For example, Chen [133–137] of Xi'an Jiaotong University produced the controllable adhesion, anisotropic superhydrophobic surfaces on PDMS by femtosecond laser scanning line by line (Fig. 1.8). This preparation method is simple and without other auxiliary steps, which can meet the demand of large area processing.

Researchers have also proposed the using of plasma etching [138, 139] of preparation of superhydrophobic surface. Although 100 nm resolution can be got through this method, it has disadvantages such as high-cost equipment, time-consuming process, and small processing area, which is difficult to meet the demand of scientific research and production.

1.4 Underwater Surperoleophobic Surfaces

In rivers and lakes which are polluted by oil spills, many water birds lose their ability to fly for their feathers are covered with oil [140, 141]. However, fish can keep their body clean in the same water. The wetting/antiwetting behavior of oil droplets on the surface of fish in water was studied and it is found that the superoleophobic fish surface originates from the water-phase micro/nanohierarchical structures and the hydrophilic chemical components of calcium phosphate, protein [140–143] (Fig. 1.9).

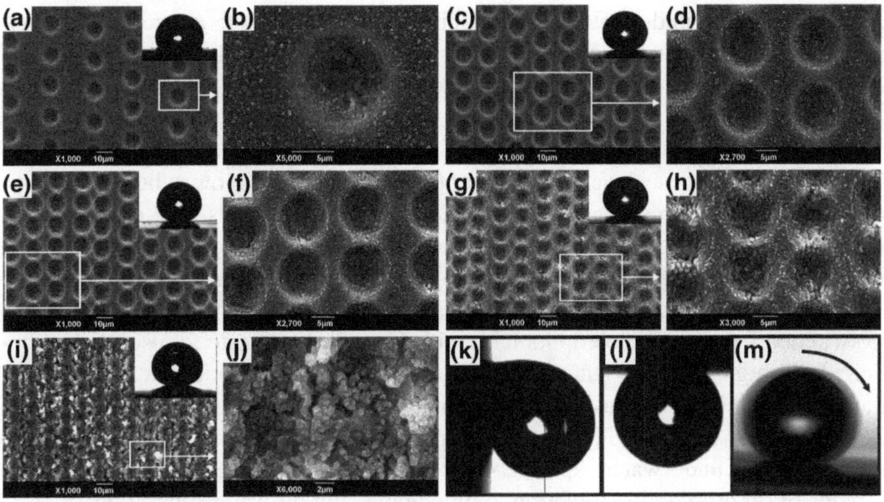

Fig. 1.8 Controllable adhesive superhydrophobic surfaces based on PDMS microwell arrays

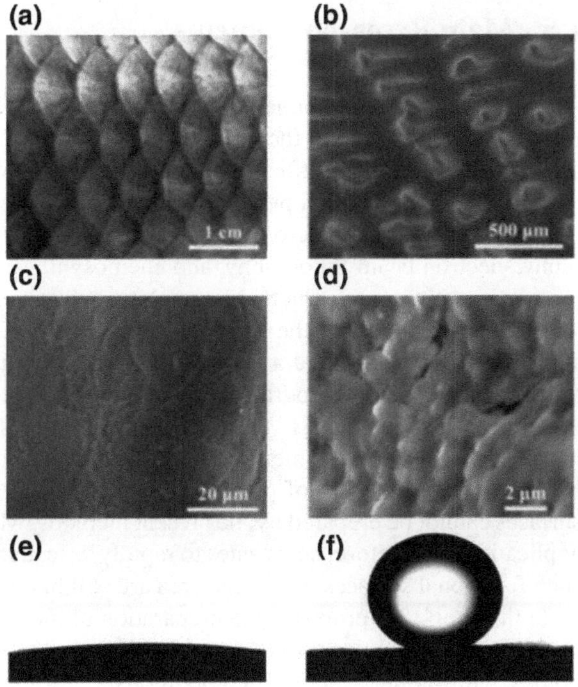

Fig. 1.9 Surface structures of fish scales and the wettability-reversion phenomena

Inspired by the fish scales, a variety of methods have been developed to prepare underwater surperoleophobic surfaces, including template method [144–148], self-assembly [149–152], electrochemical polymerization [153–155]. For example,

Li achieved robust underwater superoleophobic macromolecule-nanoclay hydrogels by replicating the hierarchical topography of fish scales on C-hydrogel surface and followed by a photo-initiated in situ radical polymerization [148]. Huang [152] presented a controllable underwater oil-adhesion-interface based on colloidal crystals assembled from nonspherical latex particles. The underwater oil-adhesive force of the as-prepared film can be effectively controlled from high to low adhesion by varying the latex structures from spherical or cauliflower-like to single cavity, which effectively adjusts the solid/liquid contact mode/wetting state of oil droplets on the films. Liu [155] synthesized superoleophobic aligned PANI nanowire films by the electrochemical polymerization on the aucoated silicon wafer. The aligned PANI can be switched from a superoleophobicity with low adhesion state to a superoleophobicity with high adhesion state by tuning an electrochemical potential. In addition, many other methods such as photolithography [156, 157], vacuum-based deposition [158, 159], sol-gel method [146], electrospinning [142, 157] have also been used to prepare various underwater superoleophobic surfaces, which have wide applications in fields of oil droplet manipulation, biomedicine, and oil/water separation.

1.5 Main Research Contents

Although various biomimetic surfaces with different functions have been successfully fabricated by taking the advantage of the advanced manufacturing equipment and methods, there are still some problems to be solved from the perspective of the scientific research and practical applications. At present, the main methods for producing biomimetic micro/nanostructures include template method, photolithography, electron beam lithography, and chemosynthesis, etc. Template method need to prepare micro/nanostructured template beforehand, and then transferred the micro/nanostructured to the target materials [116–121]. It usually takes a lot of complicated processes and a large number of templates, which is time-consuming and expensive. Photolithography and electron beam lithography need expensive instrument and apparatus [131, 132]. Chemical synthesis methods [124–130] require harsh processing environments and the chemicals may cause harm to the environment. From the view of practical applications, large area micro/nanostructures surfaces cannot be prepared by the present methods, which greatly limit the practical applications. Therefore, new routes to rapidly, effectively and economically produce such functional surfaces with large area are still highly desirable.

In this work, we proposed the preparation of bionic functional micro- and nanostructures by combining the advanced femtosecond laser micro/nanofabrication technology with bionic design. The conditions of formation of the femtosecond laser induced structures factors controlling the dimensions were in detail explored. Especially, the relevant functions were thoroughly researched.

References

1. Dickinson MH. Bionics: Biological insight into mechanical design. Proc Natl Acad Sci. 1999;96(25):14208–9.
2. Johnson EAC, Bonser RHC, Jeronimidis G. Recent advances in biomimetic sensing technologies. Philosophical transactions of the royal society of London A: Mathematical, physical and engineering sciences. 1893;2009(367):1559–69.
3. Kirchner A, Schadschneider A. Simulation of evacuation processes using a bionics-inspired cellular automaton model for pedestrian dynamics. Physica A. 2002;312(1):260–76.
4. Bjorklund B. Qualitative analysis of gel precipitates with the aid of chemical color reactions. Proc Soc Exp Biol Med. 1954;85(3):438–41.
5. Barton G M. Chemical color tests for Canadian woods. Can For Ind. 1973.
6. Takeoka Y, Watanabe M. Tuning structural color changes of porous thermosensitive gels through quantitative adjustment of the cross-linker in pre-gel solutions. Langmuir. 2003;19(22):9104–6.
7. Zhao Y, Xie Z, Gu H, et al. Bio-inspired variable structural color materials. Chem Soc Rev. 2012;41(8):3297–317.
8. Lee RT, Smith GS. Detailed electromagnetic simulation for the structural color of butterfly wings. Appl Opt. 2009;48(21):4177–90.
9. Kinoshita S, Yoshioka S. Structural colors in nature: the role of regularity and irregularity in the structure. ChemPhysChem. 2005;6(8):1442–59.
10. Greenewalt CH, Brandt W, Friel DD. Iridescent colors of hummingbird feathers. JOSA. 1960;50(10):1005–13.
11. Cong H, Cao W. Thin film interference of colloidal thin films. Langmuir. 2004;20(19):8049–53.
12. Gralak B, Tayeb G, Enoch S. Morpho butterflies wings color modeled with lamellar grating theory. Opt Express. 2001;9(11):567–78.
13. Knop K. Color pictures using the zero diffraction order of phase grating structures. Opt Commun. 1976;18(3):298–303.
14. Lochbihler H. Colored images generated by metallic sub-wavelength gratings. Opt Express. 2009;17(14):12189–96.
15. Pursiainen OLJ, Baumberg JJ, Winkler H, et al. Nanoparticle-tuned structural color from polymer opals. Opt Express. 2007;15(15):9553–61.
16. Arsenault AC, Puzzo DP, Manners I, et al. Photonic-crystal full-color displays. Nat Photonics. 2007;1(8):468–72.
17. Reeves WH, Skryabin DV, Biancalana F, et al. Transformation and control of ultra-short pulses in dispersion-engineered photonic crystal fibres. Nature. 2003;424(6948):511–5.
18. Parker AR, Townley HE. Biomimetics of photonic nanostructures. Nanosc Technol: A Collect Rev Nat J. 2010:230–6.
19. Shevtsova E, Hansson C, Janzen DH, et al. Stable structural color patterns displayed on transparent insect wings. Proc Natl Acad Sci. 2011;108(2):668–73.
20. Zi J, Yu X, Li Y, et al. Coloration strategies in peacock feathers. Proc Natl Acad Sci. 2003;100(22):12576–8.
21. Chung K, Yu S, Heo CJ, et al. Flexible, angle-independent, structural color reflectors inspired by morpho butterfly wings. Adv Mater. 2012;24(18):2375–9.
22. Kim H, Ge J, Kim J, et al. Structural color printing using a magnetically tunable and lithographically fixable photonic crystal. Nat Photonics. 2009;3(9):534–40.
23. Vorobyev AY, Guo C. Colorizing metals with femtosecond laser pulses. Appl Phys Lett. 2008;92(4):041914.
24. Vorobyev AY, Makin VS, Guo C. Brighter light sources from black metal: significant increase in emission efficiency of incandescent light sources. Phys Rev Lett. 2009;102(23):234301.
25. Vorobyev AY, Guo C. Femtosecond laser blackening of platinum. J Appl Phys. 2008;104(5):053516.

26. Vorobyev AY, Guo C. Direct creation of black silicon using femtosecond laser pulses. Appl Surf Sci. 2011;257(16):7291–4.
27. Vorobyev AY, Guo C. Reflection of femtosecond laser light in multipulse ablation of metals. J Appl Phys. 2011;110(4):043102.
28. Wu Q, Ma Y, Fang R, et al. Femtosecond laser-induced periodic surface structure on diamond film. Appl Phys Lett. 2003;82(11):1703–5.
29. Borowiec A, Haugen HK. Subwavelength ripple formation on the surfaces of compound semiconductors irradiated with femtosecond laser pulses. Appl Phys Lett. 2003;82(25):4462–4.
30. Vorobyev AY, Makin VS, Guo C. Periodic ordering of random surface nanostructures induced by femtosecond laser pulses on metals. J Appl Phys. 2007;101(3):034903.
31. Wagner R, Gottmann J, Horn A, et al. Subwavelength ripple formation induced by tightly focused femtosecond laser radiation. Appl Surf Sci. 2006;252(24):8576–9.
32. Varlamova O, Costache F, Reif J, et al. Self-organized pattern formation upon femtosecond laser ablation by circularly polarized light. Appl Surf Sci. 2006;252(13):4702–6.
33. Sano T, Yanai M, Ohmura E, et al. Femtosecond laser fabrication of microspike-arrays on tungsten surface. Appl Surf Sci. 2005;247(1):340–6.
34. Tsutsumi N, Fujihara A. Pulsed laser induced spontaneous gratings on a surface of azobenzene polymer. Appl Phys Lett. 2004;85(20):4582–4.
35. Qi L, Nishii K, Namba Y. Regular subwavelength surface structures induced by femtosecond laser pulses on stainless steel. Opt Lett. 2009;34(12):1846–8.
36. Dusser B, Sagan Z, Soder H, et al. Controlled nanostructrures formation by ultra fast laser pulses for color marking. Opt Express. 2010;18(3):2913–24.
37. Yao J, Zhang C, Liu H, et al. Selective appearance of several laser-induced periodic surface structure patterns on a metal surface using structural colors produced by femtosecond laser pulses. Appl Surf Sci. 2012;258(19):7625–32.
38. Canning J. Fibre gratings and devices for sensors and lasers. Laser Photonics Rev. 2008;2(4):275–89.
39. Li L, Hong M, Schmidt M, et al. Laser nano-manufacturing-state of the art and challenges. CIRP Ann Manuf Technol. 2011;60(2):735–55.
40. Ahsan MS, Ahmed F, Kim YG, et al. Colorizing stainless steel surface by femtosecond laser induced micro/nano-structures. Appl Surf Sci. 2011;257(17):7771–7.
41. Zhakhovskii VV, Inogamov NA, Nishihara K. New mechanism of the formation of the nanorelief on a surface irradiated by a femtosecond laser pulse. JETP Lett. 2008;87(8):423–7.
42. Ionin AA, Kudryashov SI, Makarov SV, et al. Femtosecond laser color marking of metal and semiconductor surfaces. Appl Phys A. 2012;107(2):301–5.
43. Wang X, Zhang D, Zhang H, et al. Tuning color by pore depth of metal-coated porous alumina. Nanotechnology. 2011;22(30):305306.
44. Lehmuskero A, Kontturi V, Hiltunen J, et al. Modeling of laser-colored stainless steel surfaces by color pixels. Appl Phys B: Lasers Opt. 2010;98(2):497–500.
45. Tang G, Hourd AC, Abdolvand A. Nanosecond pulsed laser blackening of copper. Appl Phys Lett. 2012;101(23):231902.
46. Li G, Li J, Yang L, et al. Evolution of aluminum surface irradiated by femtosecond laser pulses with different pulse overlaps. Appl Surf Sci. 2013;276:203–9.
47. Antończak AJ, Kocoń D, Nowak M, et al. Laser-induced colour marking-sensitivity scaling for a stainless steel. Appl Surf Sci. 2013;264:229–36.
48. Cassie ABD, Baxter S. Wettability of porous surfaces. Trans Faraday Soc. 1944;40:546–51.
49. Öner D, McCarthy TJ. Ultrahydrophobic surfaces. Effects of topography length scales on wettability. Langmuir. 2000;16(20):7777–82.
50. Sun T, Feng L, Gao X, et al. Bioinspired surfaces with special wettability. Acc Chem Res. 2005;38(8):644–52.
51. Fowkes, Frederick M, ed. Contact angle, wettability, and adhesion. Am Chem Soc. 1964.
52. Wenzel RN. Surface roughness and contact angle. J Phys Chem. 1949;53(9):1466–7.
53. Joanny JF, De Gennes PG. A model for contact angle hysteresis. Simple views on condensed matter. 2003:457–67.

54. De Gennes PG. Wetting: statics and dynamics. Rev Mod Phys. 1985;57(3):827.
55. Johnson RE Jr, Dettre RH. Contact angle hysteresis. III. Study of an idealized heterogeneous surface. J Phys Chem. 1964;68(7):1744–50.
56. Good RJ, Girifalco LA. A theory for estimation of surface and interfacial energies. III. Estimation of surface energies of solids from contact angle data. J Phys Chem. 1960;64(5):561–5.
57. Dettre RH, Johnson RE Jr. Contact Angle Hysteresis. IV. Contact angle measurements on heterogeneous surfaces. J Phys Chem. 1965;69(5):1507–15.
58. Brackbill JU, Kothe DB, Zemach C. A continuum method for modeling surface tension. J Comput Phys. 1992;100(2):335–54.
59. Tyson WR, Miller WA. Surface free energies of solid metals: Estimation from liquid surface tension measurements. Surf Sci. 1977;62(1):267–76.
60. Tolman RC. The effect of droplet size on surface tension. J Chem Phys. 1949;17(3):333–7.
61. Rotenberg Y, Boruvka L, Neumann AW. Determination of surface tension and contact angle from the shapes of axisymmetric fluid interfaces. J Colloid Interface Sci. 1983;93(1):169–83.
62. Kirkwood JG, Buff FP. The statistical mechanical theory of surface tension. J Chem Phys. 1949;17(3):338–43.
63. Van Oss CJ, Good RJ, Chaudhury MK. Additive and nonadditive surface tension components and the interpretation of contact angles. Langmuir. 1988;4(4):884–91.
64. Jasper JJ. The surface tension of pure liquid compounds. J Phys Chem Ref Data. 1972;1(4):841–1010.
65. Wenzel RN. Resistance of solid surfaces to wetting by water. Ind Eng Chem. 1936;28(8):988–94.
66. Zhu X, Zhu L, Chen H, et al. Micro-ball lens structure fabrication based on drop on demand printing the liquid mold. Appl Surf Sci. 2016;361:80–9.
67. Good RJ. A thermodynamic derivation of wenzel's modification of young's equation for contact angles; together with a theory of hysteresis. J Am Chem Soc. 1952;74(20):5041–2.
68. Pierce E, Carmona FJ, Amirfazli A. Understanding of sliding and contact angle results in tilted plate experiments. Colloids Surf, A. 2008;323(1):73–82.
69. Nakajima A, Abe K, Hashimoto K, et al. Preparation of hard super-hydrophobic films with visible light transmission. Thin Solid Films. 2000;376(1):140–3.
70. Yoshimitsu Z, Nakajima A, Watanabe T, et al. Effects of surface structure on the hydrophobicity and sliding behavior of water droplets. Langmuir. 2002;18(15):5818–22.
71. Miwa M, Nakajima A, Fujishima A, et al. Effects of the surface roughness on sliding angles of water droplets on superhydrophobic surfaces. Langmuir. 2000;16(13):5754–60.
72. Kawasaki K. Study of wettability of polymers by sliding of water drop. J Colloid Sci. 1960;15(5):402–7.
73. Fürstner R, Barthlott W, Neinhuis C, et al. Wetting and self-cleaning properties of artificial superhydrophobic surfaces. Langmuir. 2005;21(3):956–61.
74. Xie Q, Xu J, Feng L, et al. Facile Creation of a super-amphiphobic coating surface with bionic microstructure. Adv Mater. 2004;16(4):302–5.
75. Wang S, Feng L, Jiang L. One-step solution-immersion process for the fabrication of stable bionic superhydrophobic surfaces. Adv Mater. 2006;18(6):767–70.
76. Artus GRJ, Jung S, Zimmermann J, et al. Silicone nanofilaments and their application as superhydrophobic coatings. Adv Mater. 2006;18(20):2758–62.
77. Choi SJ, Suh KY, Lee HH. A geometry controllable approach for the fabrication of biomimetic hierarchical structure and its superhydrophobicity with near-zero sliding angle. Nanotechnology. 2008;19(27):275305.
78. Guo Z, Zhou F, Hao J, et al. Stable biomimetic super-hydrophobic engineering materials. J Am Chem Soc. 2005;127(45):15670–1.
79. Kiuru M, Alakoski E. Low sliding angles in hydrophobic and oleophobic coatings prepared with plasma discharge method. Mater Lett. 2004;58(16):2213–6.
80. Acatay K, Simsek E, Ow-Yang C, et al. Tunable, superhydrophobically stable polymeric surfaces by electrospinning. Angew Chem Int Ed. 2004;43(39):5210–3.

81. White AM, Truesdale MC, Bae JG, et al. Differential effects of ethanol on motor coordination in adolescent and adult rats. Pharmacol Biochem Behav. 2002;73(3):673–7.
82. Langmuir I. The mechanism of the surface phenomena of flotation. Trans Faraday Soc. 1920;15(June):62–74.
83. Suzuki S, Nakajima A, Kameshima Y, et al. Elongation and contraction of water droplet during sliding on the silicon surface treated by fluoroalkylsilane. Surf Sci. 2004;557(1):L163–8.
84. Kamitani K, Teranishi T. Development of water-repellent glass improved water-sliding property and durability. J Sol-Gel Sci Technol. 2003;26(1):823–5.
85. Sakai M, Song JH, Yoshida N, et al. Direct observation of internal fluidity in a water droplet during sliding on hydrophobic surfaces. Langmuir. 2006;22(11):4906–9.
86. Erbil HY, Demirel AL, Avcı Y, et al. Transformation of a simple plastic into a superhydrophobic surface. Science. 2003;299(5611):1377–80.
87. Jiang L, Zhao Y, Zhai J. A lotus-leaf-like superhydrophobic surface: a porous microsphere/nanofiber composite film prepared by electrohydrodynamics. Angew Chem. 2004;116(33):4438–41.
88. Roach P, Shirtcliffe NJ, Newton MI. Progess in superhydrophobic surface development. Soft Matter. 2008;4(2):224–40.
89. Li XM, Reinhoudt D, Crego-Calama M. What do we need for a superhydrophobic surface? A review on the recent progress in the preparation of superhydrophobic surfaces. Chem Soc Rev. 2007;36(8):1350–68.
90. Feng L, Song Y, Zhai J, et al. Creation of a superhydrophobic surface from an amphiphilic polymer. Angew Chem. 2003;115(7):824–6.
91. Choi CH, Kim CJ. Large slip of aqueous liquid flow over a nanoengineered superhydrophobic surface. Phys Rev Lett. 2006;96(6):066001.
92. Huang L, Lau SP, Yang HY, et al. Stable superhydrophobic surface via carbon nanotubes coated with a ZnO thin film. J Phys Chem B. 2005;109(16):7746–8.
93. Hong X, Gao X, Jiang L. Application of superhydrophobic surface with high adhesive force in no lost transport of superparamagnetic microdroplet. J Am Chem Soc. 2007;129(6):1478–9.
94. Han JT, Lee DH, Ryu CY, et al. Fabrication of superhydrophobic surface from a supramolecular organosilane with quadruple hydrogen bonding. J Am Chem Soc. 2004;126(15):4796–7.
95. Barthlott W, Neinhuis C. Purity of the sacred lotus, or escape from contamination in biological surfaces. Planta. 1997;202(1):1–8.
96. Neinhuis C, Barthlott W. Characterization and distribution of water-repellent, self-cleaning plant surfaces. Ann Bot. 1997;79(6):667–77.
97. Fürstner R, Barthlott W, Neinhuis C, et al. Wetting and self-cleaning properties of artificial superhydrophobic surfaces. Langmuir. 2005;21(3):956–61.
98. Wagner T, Neinhuis C, Barthlott W. Wettability and contaminability of insect wings as a function of their surface sculptures. Acta Zoologica. 1996;77(3):213–25.
99. Wagner P, Fürstner R, Barthlott W, et al. Quantitative assessment to the structural basis of water repellency in natural and technical surfaces. J Exp Bot. 2003;54(385):1295–303.
100. Neinhuis C, Barthlott W. Seasonal changes of leaf surface contamination in beech, oak, and ginkgo in relation to leaf micromorphology and wettability. New Phytol. 1998;138(1):91–8.
101. Zheng Y, Han D, Zhai J, et al. In situ investigation on dynamic suspending of microdroplet on lotus leaf and gradient of wettable micro-and nanostructure from water condensation. Appl Phys Lett. 2008;92(8):084106.
102. Zhang J, Sheng X, Jiang L. The dewetting properties of lotus leaves. Langmuir. 2008;25(3):1371–6.
103. Cheng Q, Li M, Zheng Y, et al. Janus interface materials: superhydrophobic air/solid interface and superoleophobic water/solid interface inspired by a lotus leaf. Soft Matter. 2011;7(13):5948–51.
104. Liu K, Zhang M, Zhai J, et al. Bioinspired construction of Mg-Li alloys surfaces with stable superhydrophobicity and improved corrosion resistance. Appl Phys Lett. 2008;92(18):183103.

105. Sun T, Feng L, Gao X, et al. Bioinspired surfaces with special wettability. Acc Chem Res. 2005;38(8):644–52.
106. Feng XQ, Gao X, Wu Z, et al. Superior water repellency of water strider legs with hierarchical structures: experiments and analysis. Langmuir. 2007;23(9):4892–6.
107. Jiang L, Yao X, Li H, et al. "Water strider" legs with a self-assembled coating of single-crystalline nanowires of an organic semiconductor. Adv Mater. 2010;22(3):376–9.
108. Ding Y, Xu S, Zhang Y, et al. Modifying the anti-wetting property of butterfly wings and water strider legs by atomic layer deposition coating: surface materials versus geometry. Nanotechnology. 2008;19(35):355708.
109. Shi F, Niu J, Liu J, et al. Towards understanding why a superhydrophobic coating is needed by water striders. Adv Mater. 2007;19(17):2257–61.
110. Shi F, Wang Z, Zhang X. Combining a layer-by-layer assembling technique with electrochemical deposition of gold aggregates to mimic the legs of water striders. Adv Mater. 2005;17(8):1005–9.
111. Wei PJ, Chen SC, Lin JF. Adhesion forces and contact angles of water strider legs. Langmuir. 2008;25(3):1526–8.
112. Ji XY, Wang JW, Feng XQ. Role of flexibility in the water repellency of water strider legs: theory and experiment. Phys Rev E. 2012;85(2):021607.
113. Watson GS, Cribb BW, Watson JA. Experimental determination of the efficiency of nanostructuring on non-wetting legs of the water strider. Acta Biomater. 2010;6(10):4060–4.
114. Zheng QS, Yu Y, Feng XQ. The role of adaptive-deformation of water strider leg in its walking on water. J Adhes Sci Technol. 2009;23(3):493–501.
115. Pan Q, Liu J, Zhu Q. A water strider-like model with large and stable loading capacity fabricated from superhydrophobic copper foils. ACS Appl Mater Interfaces. 2010;2(7):2026–30.
116. Li Y, Cai W, Duan G, et al. Superhydrophobicity of 2D ZnO ordered pore arrays formed by solution-dipping template method. J Colloid Interface Sci. 2005;287(2):634–9.
117. Zhu Y, Hu D, Wan MX, et al. Conducting and superhydrophobic rambutan-like hollow spheres of polyaniline. Adv Mater. 2007;19(16):2092–6.
118. Wan M. A template-free method towards conducting polymer nanostructures. Adv Mater. 2008;20(15):2926–32.
119. Bormashenko E, Stein T, Whyman G, et al. Wetting properties of the multiscaled nanostructured polymer and metallic superhydrophobic surfaces. Langmuir. 2006;22(24):9982–5.
120. Feng XJ, Jiang L. Design and creation of superwetting/antiwetting surfaces. Adv Mater. 2006;18(23):3063–78.
121. Li Y, Jia WZ, Song YY, et al. Superhydrophobicity of 3D porous copper films prepared using the hydrogen bubble dynamic template. Chem Mater. 2007;19(23):5758–64.
122. Sun M, Luo C, Xu L, et al. Artificial lotus leaf by nanocasting. Langmuir. 2005;21(19):8978–81.
123. Lee Y, Park SH, Kim KB, et al. Fabrication of hierarchical structures on a polymer surface to mimic natural superhydrophobic surfaces. Adv Mater. 2007;19(17):2330–5.
124. Chen H, Wang N, Di J, et al. Nanowire-in-microtube structured core/shell fibers via multifluidic coaxial electrospinning. Langmuir. 2010;26(13):11291–6.
125. Zhao N, Shi F, Wang Z, et al. Combining layer-by-layer assembly with electrodeposition of silver aggregates for fabricating superhydrophobic surfaces. Langmuir. 2005;21(10):4713–6.
126. Tang Z, Wang Y, Podsiadlo P, et al. Biomedical applications of layer-by-layer assembly: from biomimetics to tissue engineering. Adv Mater. 2006;18(24):3203–24.
127. Kotov NA. Layer-by-layer self-assembly: the contribution of hydrophobic interactions. Nanostruct Mater. 1999;12(5):789–96.
128. Zhao Y, Li M, Lu Q, et al. Superhydrophobic polyimide films with a hierarchical topography: combined replica molding and layer-by-layer assembly. Langmuir. 2008;24(21):12651–7.
129. Serizawa T, Kamimura S, Kawanishi N, et al. Layer-by-layer assembly of poly (vinyl alcohol) and hydrophobic polymers based on their physical adsorption on surfaces. Langmuir. 2002;18(22):8381–5.

130. Amigoni S, Taffin de Givenchy E, Dufay M, et al. Covalent layer-by-layer assembled superhydrophobic organic-inorganic hybrid films. Langmuir. 2009;25(18):11073–7.
131. Khorasani MT, Mirzadeh H. In vitro blood compatibility of modified PDMS surfaces as superhydrophobic and superhydrophilic materials. J Appl Polym Sci. 2004;91(3):2042–7.
132. Long J, Fan P, Zhong M, et al. Superhydrophobic and colorful copper surfaces fabricated by picosecond laser induced periodic nanostructures. Appl Surf Sci. 2014;311:461–7.
133. Yong J, Chen F, Yang Q, et al. Controllable adhesive superhydrophobic surfaces based on PDMS microwell arrays. Langmuir. 2013;29(10):3274–9.
134. Yong J, Chen F, Yang Q, et al. Femtosecond laser weaving superhydrophobic patterned PDMS surfaces with tunable adhesion. J Phys Chem C. 2013;117(47):24907–12.
135. Yong J, Yang Q, Chen F, et al. Superhydrophobic PDMS surfaces with three-dimensional (3D) pattern-dependent controllable adhesion. Appl Surf Sci. 2014;288:579–83.
136. Yong J, Yang Q, Chen F, et al. A simple way to achieve superhydrophobicity, controllable water adhesion, anisotropic sliding, and anisotropic wetting based on femtosecond-laser-induced line-patterned surfaces. J Mater Chem A. 2014;2(15):5499–507.
137. Yong J, Yang Q, Chen F, et al. Bioinspired superhydrophobic surfaces with directional adhesion. RSC Adv. 2014;4(16):8138–43.
138. Balu B, Breedveld V, Hess DW. Fabrication of "roll-off" and "sticky" superhydrophobic cellulose surfaces via plasma processing. Langmuir. 2008;24(9):4785–90.
139. Balu B, Kim JS, Breedveld V, et al. Tunability of the adhesion of water drops on a superhydrophobic paper surface via selective plasma etching. J Adhes Sci Technol. 2009;23(2):361–80.
140. Liu M, Wang S, Wei Z, et al. Bioinspired design of a superoleophobic and low adhesive water/solid interface. Adv Mater. 2009;21(6):665–9.
141. Liu K, Tian Y, Jiang L. Bio-inspired superoleophobic and smart materials: design, fabrication, and application. Prog Mater Sci. 2013;58(4):503–64.
142. Xue Z, Jiang L. Bioinspired underwater superoleophobic surfaces. Acta Polym Sin. 2012;10:1091–101.
143. Liu X, Gao J, Xue Z, et al. Bioinspired oil strider floating at the oil/water interface supported by huge superoleophobic force. ACS Nano. 2012;6(6):5614–20.
144. Cheng Q, Li M, Zheng Y, et al. Janus interface materials: superhydrophobic air/solid interface and superoleophobic water/solid interface inspired by a lotus leaf. Soft Matter. 2011;7(13):5948–51.
145. Yao X, Song Y, Jiang L. Applications of bioinspired special wettable surfaces. Adv Mater. 2011;23(6):719–34.
146. Sawai Y, Nishimoto S, Kameshima Y, et al. Photoinduced underwater superoleophobicity of TiO_2 thin films. Langmuir. 2013;29(23):6784–9.
147. Tian Y, Jiang L. Design of bioinspired, smart, multiscale interfacial materials with superwettability. MRS Bull. 2015;40(2):155–65.
148. Lin L, Liu M, Chen L, et al. Bio-inspired hierarchical macromolecule-nanoclay hydrogels for robust underwater superoleophobicity. Adv Mater. 2010;22(43):4826–30.
149. Cao Y, Zhang X, Tao L, et al. Mussel-inspired chemistry and michael addition reaction for efficient oil/water separation. ACS Appl Mater Interfaces. 2013;5(10):4438–42.
150. Zhou X, Zhang Z, Xu X, et al. Robust and durable superhydrophobic cotton fabrics for oil/water separation. ACS Appl Mater Interfaces. 2013;5(15):7208–14.
151. Song J, Liu H, Wan M, et al. Bio-inspired isotropic and anisotropic wettability on a Janus free-standing polypyrrole film fabricated by interfacial electro-polymerization. J Mater Chem A. 2013;1(5):1740–4.
152. Huang Y, Liu M, Wang J, et al. Controllable underwater oil-adhesion-interface films assembled from nonspherical particles. Adv Func Mater. 2011;21(23):4436–41.
153. Cheng Q, Li M, Yang F, et al. An underwater pH-responsive superoleophobic surface with reversibly switchable oil-adhesion. Soft Matter. 2012;8(25):6740–3.
154. Taleb S, Darmanin T, Guittard F. Elaboration of voltage and ion exchange stimuli-responsive conducting polymers with selective switchable liquid-repellency. ACS Appl Mater Interfaces. 2014;6(10):7953–60.

155. Ding C, Zhu Y, Liu M, et al. PANI nanowire film with underwater superoleophobicity and potential-modulated tunable adhesion for no loss oil droplet transport. Soft Matter. 2012;8(35):9064–8.
156. Wu D, Wu S, Chen QD, et al. Facile creation of hierarchical PDMS microstructures with extreme underwater superoleophobicity for anti-oil application in microfluidic channels. Lab Chip. 2011;11(22):3873–9.
157. Jiang T, Guo Z, Liu W. Biomimetic superoleophobic surfaces: focusing on their fabrication and applications. J Mater Chem A. 2015;3(5):1811–27.
158. Jin H, Kettunen M, Laiho A, et al. Superhydrophobic and superoleophobic nanocellulose aerogel membranes as bioinspired cargo carriers on water and oil. Langmuir. 2011;27(5):1930–4.
159. Liu K, Jiang L. Bio-inspired self-cleaning surfaces. Annu Rev Mater Res. 2012;42:231–63.

Chapter 2
Key Technical Problems of Femtosecond Laser Bionic Surfaces

2.1 Femtosecond Laser Micro/Nanofabrication Equipment and Systems

Researchers of micro/nano engineering laboratory of University of Science and Technology of China have established the femtosecond laser micro/nanofabrication system since 2001, as shown in Fig. 2.1. This micro/nanofabrication system is composed of femtosecond laser system, precision processing system, and control system.

Basic components and parameters:

(1) Femtosecond laser system: This system consists of seed laser (Chameleon VISION-S, coherent), titanium sapphire regeneration amplification system (Legend Elite-1 K-HE, coherent), and optical parametric amplifier (coherent). The output laser is linearly polarized with the pulse repetition frequency of 1 kHz, the pulse width of 104 fs, and center wavelength of 800 nm. The laser pulse energy can be adjusted from 0 to 4.5 mJ. The wavelength can be adjusted from 220 to 2600 nm by the transformation of the optical parametric amplifier.
(2) Precision processing system: This part is composed of high precision *XY* scanning galvanometer (SAMLight, Germany) and *Z*-direction mobile platform. The focal length achromatic focusing lens is 63 mm and the size of the focal point is about 22 μm. The laser beams can be scanned in the *X* and Y directions, and form a circular region with diameter of 45 mm. With the help of *Z*-direction mobile platform whose work distance is 10 mm, the sample can be precisely moved to the location of the focal point [1].
(3) Control system: The function of this system is to realize the recognition and communication between equipment.

The optical pathway diagram of the femtosecond laser micro/nanofabrication system is shown in Fig. 2.2. The pulse energy could be continuously varied by using the combination of a $\lambda/2$ wave plate and a Glan-Taylor polarizer. The average energy of the laser pulse is measured by a power meter (Coherent, FM10), which is set behind the polarizer.

G. Li, *Bionic Functional Structures by Femtosecond Laser Micro/nanofabrication Technologies*, Springer Theses, https://doi.org/10.1007/978-981-13-0359-3_2

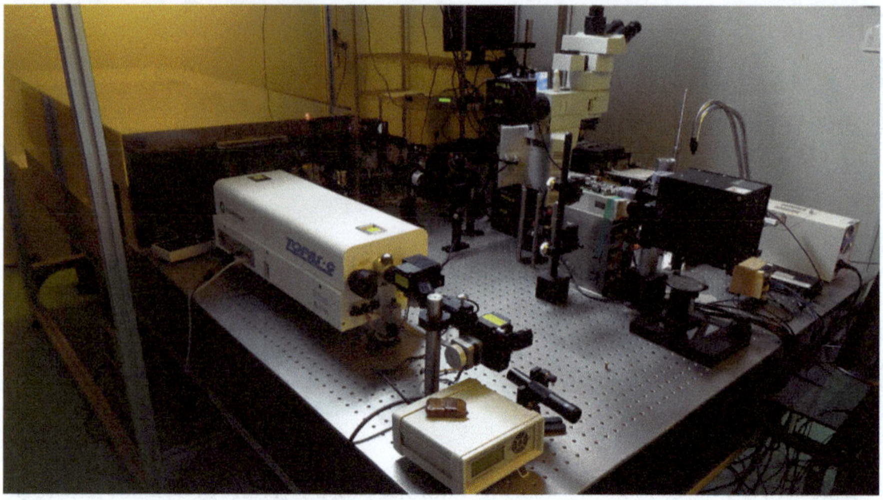

Fig. 2.1 Photograph of femtosecond laser micro/nanofabrication equipment and systems in our laboratory

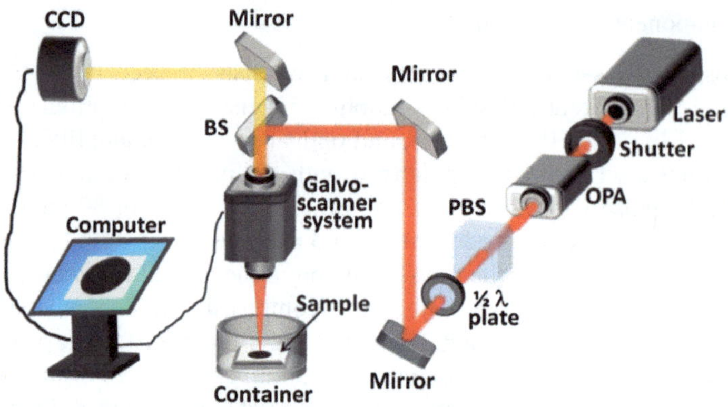

Fig. 2.2 Schematic representation of femtosecond laser micro/nanofabrication system

2.2 Material and Their Parameters

In the experiment of femtosecond laser micro/nanofabrication, the main used materials are metal and polymer. All the metals are purchased from New Metal Material Tech. Co., Ltd, Beijing, China, and the detailed parameters are shown in Table 2.1.

The polymer is the Polydimethylsiloxane (PDMS), which is optically clear, and, in general, inert, non-toxic, non-flammable, and is widely used in soft lithography, microfluidics, etc. The liquid environment where femtosecond laser beam irradiates the sample is water, ethanol, and sucrose solution. The water is homemade distilled water. The ethanol whose purity is 99.3%, is purchased from Sinopharm chemical

Table 2.1 Material and their parameters

Name	Purity	Size
Aluminum	99.99%	$10 \times 10 \times 0.2$ mm^3, $20 \times 20 \times 0.2$ mm^3
Stainless steel (316)		$30 \times 20 \times 0.2$ mm^3
Nickel	99.99%	$20 \times 20 \times 0.2$ mm^3
Silicon		$20 \times 20 \times 0.2$ mm^3, $20 \times 40 \times 0.2$ mm^3

reagent Co., Ltd. The sucrose solution was made by dissolving a certain amount of sucrose into 100 g distilled water. In wettability test, the oil is CH_2ClCH_2Cl, and the molecular weight is 98.96, which is also purchased from Sinopharm chemical reagent Co., Ltd. The red ink, and the milk are purchased from nearby store.

2.3 Technological Process

The steps for preparing the bionic surfaces can be summarized as follows:

(1) The sample surface was cleaned with acetone, ethanol, and distilled water for 10 min and then dried in nitrogen.
(2) Placing the sample onto the *Z*-direction mobile platform and then focusing the laser beam onto the surface.
(3) Load the pre-processing target such as the word, figure, pattern to the control system.
(4) Setting the appropriate pulse energy, scanning speed, scanning interval etc. according to the experimental requirements.
(5) Writing the target objects onto the sample surface by scanning the focused laser beam line by line.

If the experiment is to be conducted with the assist of liquid environment, the sample is beforehand fixed at the bottom of the container followed by adding liquid.

2.4 The Influence of Laser Parameters and Environment on the Morphology of the Micro/Nanostructures

In the experiment of femtosecond laser micro/nanofabrication, the laser processing parameters have a decisive influence on the morphology of the structures. These parameters include laser scanning speed [1–3], pulse energy [3], scanning interval, polarization direction, and processing environment [4–11] and so on.

Firstly, the influence of laser scanning speed and pulse energy on the morphology on stainless steel surface in air is studied.

It is indicated that at scanning speed of 12 mm/s, the femtosecond laser can induce subwavelength periodic ripples in large energy range. Here the energy range from 0.05 to 0.20 mJ is chosen to study the influence of pulse energy on the morphology of the periodic ripples, and the results is shown in Fig. 2.3. At pulse energy of 0.05 mJ, the periodic ripples with period of 520 nm can be induced (Fig. 2.3a). However, the ripples are uneven, and even some place of the stainless steel surface is not covered with ripples. When the pulse energy is increased to 0.10 mJ, it is found that smooth, a small amount of granular structure covered ripples with period of 560 nm are formed (Fig. 2.3b). Compared to the results in Fig. 2.3a, the ripples formed at pulse energy of 0.10 mJ are long and straight, and there is no evident fragmentation. With the pulse energy increased to 0.15 mJ, the previous long and straight ripples become bent, and there are a large number of spherical structures with size from 50 to 300 nm formed on or between the ripples Fig. 2.3c). Although the period of the ripples can be further increased to 630 nm with the pulse energy increased to 0.20 mJ, many long cracks appears Fig. 2.3d). From the amplified SEM images in Fig. 2.4, it is clear that the cracks tear the original whole ripples into two small ripples with period of about 350 nm. In addition, the intervals between the adjacent ripples formed at 0.20 mJ are about 160–200 nm, which are evidently larger than that of the small pulse energy. Furthermore, many dozens of nanoscale filaments form on the ripples surfaces or bridge the adjacent ripples with the orientation perpendicular to the ripples. It is considered that the formation mechanism is attributed to the gaussian spatial distribution of pulse energy [12–16].

Sakabe [17, 18] proposed that the ripples is formed due to the superposition of the multipulse energy on the metal surface, and the overall process can be summarized as follows: (1) plasma waves are induced on metal surface by the femtosecond laser pulse; (2) spatially localized ion clouds Coulomb-explode to vacuum, and consequently the thin layer is ablated, and the interspacing of the gratings is printed at the stage of the first several pulses; (3) for the subsequent pulses, the electric field is enhanced near the initially printed structures, and the near field ablates the surface, thus deepening the structures. In addition, it is thought that the period (d) of the ripples is approximately equal to the wavelength of laser induced plasma waves (λ_{SP}). Hence, the formula for the period can be summarized as [17, 18]:

$$d = \lambda_{SP} = \lambda_L \left\{ 1 + \left(\frac{n_{es} \cdot e^2 \cdot \lambda_L^2}{mc^2} - 2 \right)^{-1} \right\}^{-1/2} \tag{2.1}$$

where λ_L is the incident laser wavelength, n_{es} is the surface electron density induced by the incident laser, c is the speed of light in vacuum, e is the electron charge, and m is the electron mass, respectively. n_{es} in our case is influenced by the laser fluence and the incident laser wavelength. From Eq. 2.1, an obvious and direct correspondence can be obtained between ripple periods d and incident laser wavelength λ_L. In addition, according to the literature, the period of ripples changes from 0.5 λ_L to 0.85 λ_L when the incident laser wavelength of 800 nm, which is agreed with the experimental results.

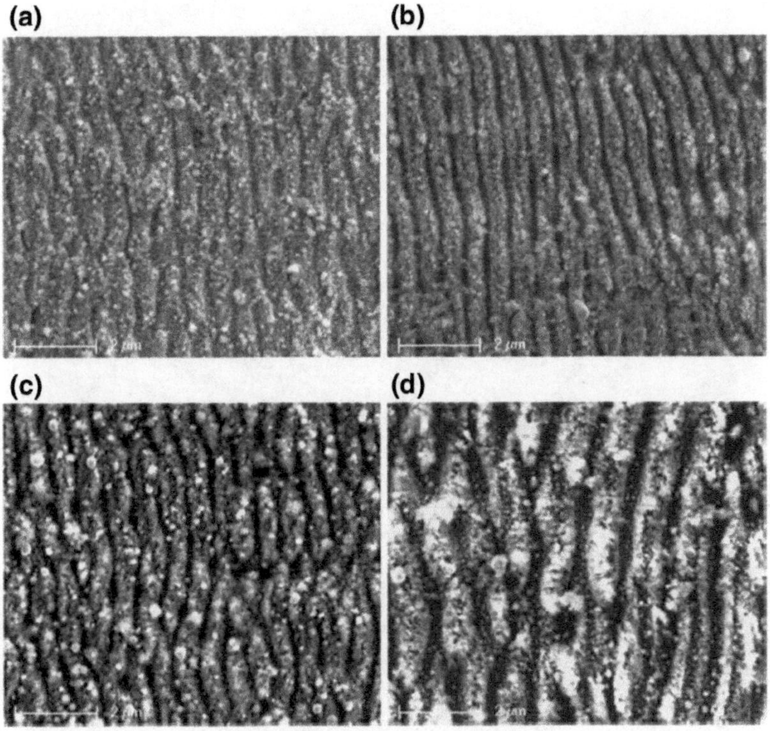

Fig. 2.3 The influence of laser pulse energy on the morphology

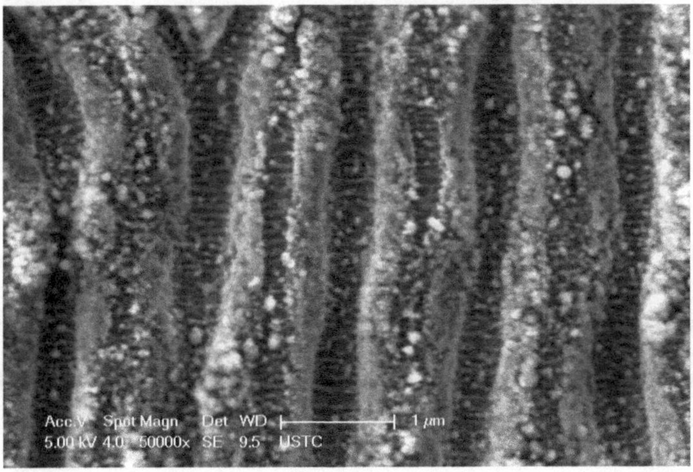

Fig. 2.4 Amplified SEM images of lacerated ripples

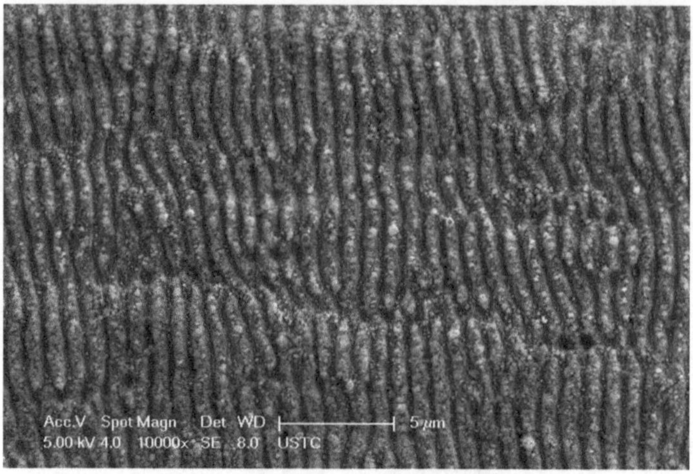

Fig. 2.5 High-quality long and straight ripples with period of 540 nm

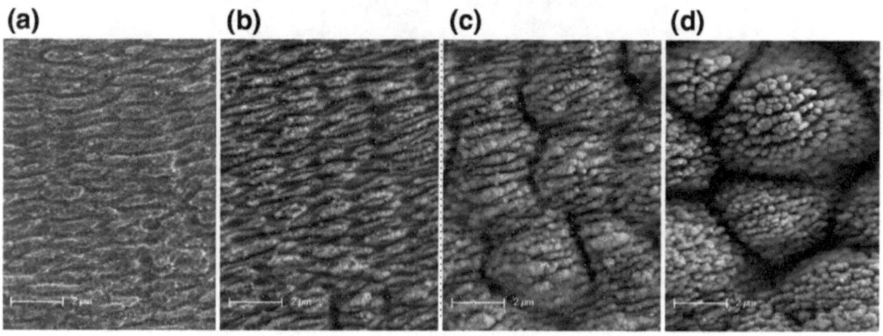

Fig. 2.6 SEM images of structures of stainless steel surface produced at scanning speed of 15, 10, 5, and 1 mm/s with pulse energy of 0.10 mJ

Based on the above discussion, at pulse energy of 0.092 mJ and scanning speed of 12 mm/s, high-quality long and straight ripples with period of 540 nm are produced, and the SEM image is shown in Fig. 2.5. Besides, the morphology of the structures is also really relevant to the laser scanning speed [1]. Figure 2.6 shows the SEM images of structures of stainless steel surface produced at scanning speed of 15, 10, 5, and 1 mm/s with pulse energy of 0.10 mJ.

The uniform ripple with period of about 550 nm can be induced when the scanning speed is 15 mm/s (Fig. 2.6a). When the scanning speed is decreased to 10 mm/s, it is found that the long and straight ripples are fractured to the shorter ones with length of 3–7 μm (Fig. 2.6b). Further decreasing the scanning speed to 5 mm/s, most ripples have disappeared meanwhile, blocks of parallel nanoscale lines are formed in the original region (Fig. 2.6c). In addition, the directions of the nanoscale lines are consistent with that of the ripples. However, the ripples completely disappeared,

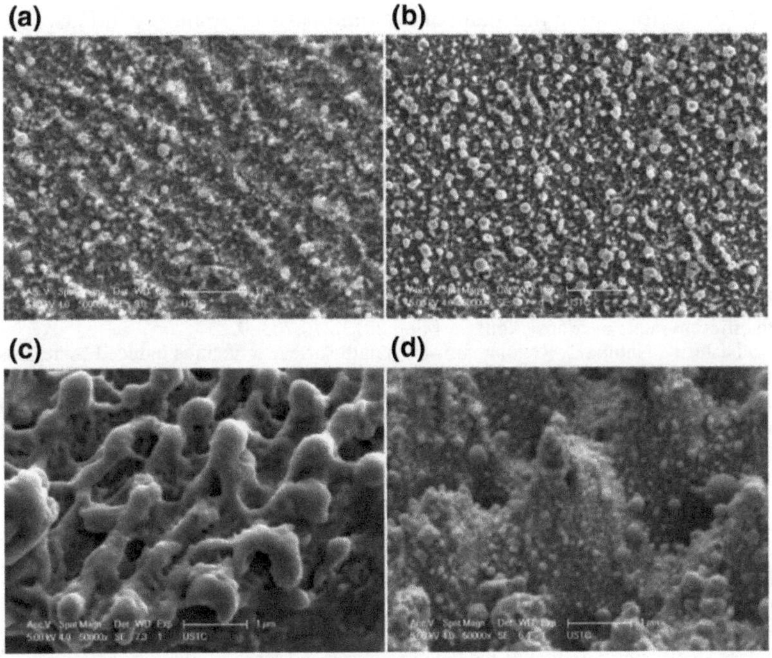

Fig. 2.7 The influence of liquid environment on the surface morphology

and instead appeared conical structures with diameter of 2–5 μm. Furthermore, the micro-scale conical structures are covered with nanoparticles, looking like the strawberries (Fig. 2.6d).

Recently, the liquid- assisted femtosecond laser micro/nanofabrication has become more and more popular due to the efficient cooling effect, high pressure of plasma, and pollution-free of modified surface [4–9, 11]. Figure 2.7 shows the nickel surface structures prepared in air (Fig. 2.7a), water (Fig. 2.7b), ethanol (Fig. 2.7c), and sucrose solution (Fig. 2.7d), respectively at pulse energy of 0.10 mJ at scanning speed of 2 mm/s. It is observed that completely different structures can be formed in different environments. Although the formation mechanism of surface structures by femtosecond laser irradiation in air has been deeply studied, that in liquid environment is still not clear. It is considered that the structures formed in different liquids are related to the different physical and chemical effects. The plasma is formed on the liquid/solid interface when the laser beam transmitted through the liquid layer to the material surface. The evolution and formation of the structures involve complex laser-liquid-material mechanisms, such as interference of incident laser and plasma waves, liquid-induced uniform distribution of pulse energy, prolonged melting and resolidification process, and mechanical pressure remodeling process [4–9, 11].

Although the formation mechanism of liquid-assisted femtosecond laser induced micro/nanostructures is not clear, we have understood the morphological characteristics of the structures and the evolution, which provide the way to accurately control the structures.

References

1. Li G, Li J, Yang L, et al. Evolution of aluminum surface irradiated by femtosecond laser pulses with different pulse overlaps. Appl Surf Sci. 2013;276:203–9.
2. Qi L, Nishii K, Namba Y. Regular subwavelength surface structures induced by femtosecond laser pulses on stainless steel. Opt Lett. 2009;34(12):1846–8.
3. Yao J, Zhang C, Liu H, et al. Selective appearance of several laser-induced periodic surface structure patterns on a metal surface using structural colors produced by femtosecond laser pulses. Appl Surf Sci. 2012;258(19):7625–32.
4. Yang HD, Li XH, Li GQ, et al. Formation of colorized silicon by femtosecond laser pulses in different background gases. Appl Phys A Mater Sci Process. 2011;104(2):749–53.
5. Amoruso S, Bruzzese R, Wang X, et al. Femtosecond laser ablation of nickel in vacuum. J Phys D Appl Phys. 2007;40(2):331.
6. Ali N, Bashir S, Akram M, et al. Effect of dry and wet ambient environment on the pulsed laser ablation of titanium. Appl Surf Sci. 2013;270:49–57.
7. Bian H, Yang Q, Liu H, et al. A facile preparation route for netlike microstructures on a stainless steel using an ethanol-mediated femtosecond laser irradiation. Mater Sci Eng, C. 2013;33(2):663–7.
8. Barmina EV, Stratakis E, Fotakis K, et al. Generation of nanostructures on metals by laser ablation in liquids: new results. Quantum Electron. 2010;40(11):1012.
9. Kumar B, Yadav D, Thareja RK. Growth dynamics of nanoparticles in laser produced plasma in liquid ambient. J Appl Phys. 2011;110(7):074903.
10. Amoruso S, Ausanio G, Barone AC, et al. Nanoparticles size modifications during femtosecond laser ablation of nickel in vacuum. Appl Surf Sci. 2007;254(4):1012–6.
11. Albu C, Dinescu A, Filipescu M, et al. Periodical structures induced by femtosecond laser on metals in air and liquid environments. Appl Surf Sci. 2013;278:347–51.
12. Bonse J, Rosenfeld A, Krüger J. Femtosecond laser-induced periodic surface structures: recent approaches to explain their sub-wavelength periodicities. In: LAT 2010: International conference on lasers, applications, and technologies. International society for optics and photonics; 2011, 7994, 79940M.
13. Le Harzic R, Dörr D, Sauer D, et al. Generation of high spatial frequency ripples on silicon under ultrashort laser pulses irradiation. Appl Phys Lett. 2011;98(21):211905.
14. Varlamova O, Bounhalli M, Reif J. Influence of irradiation dose on laser-induced surface nanostructures on silicon. Appl Surf Sci. 2013;278:62–6.
15. Obona JV, Ocelík V, Skolski JZP, et al. On the surface topography of ultrashort laser pulse treated steel surfaces. Appl Surf Sci. 2011;258(4):1555–60.
16. Straub M, Afshar M, Feili D, et al. Surface plasmon polariton model of high-spatial frequency laser-induced periodic surface structure generation in silicon. J Appl Phys. 2012;111(12):124315.
17. Okamuro K, Hashida M, Miyasaka Y, et al. Laser fluence dependence of periodic grating structures formed on metal surfaces under femtosecond laser pulse irradiation. Phys Rev B. 2010;82(16):165417.
18. Sakabe S, Hashida M, Tokita S, et al. Mechanism for self-formation of periodic grating structures on a metal surface by a femtosecond laser pulse. Phys Rev B. 2009;79(3):033409.

Chapter 3
Study on Femtosecond Laser Induced Structural Colors

In this chapter, we prepared various micro/nanostructures and realized the colorizing the materials surface. By adopting different laser pulse overlaps several kinds of structural colors were produced on the aluminum surfaces. In addition, we not only discussed the combined influence of incident light angle and the ripples orientation on the diversity of structural colors, but also proposed the possibility of producing multi-patterns constituted by ripples with different orientations, and realized the diverse and exclusive display. Furthermore, the preparation of three-dimensional effect patterns on metal surface was proposed.

3.1 Femtosecond Laser Induced Structural Colors on Aluminum Surfaces

Aluminum is a common energetic material, and it has the characteristics of low melting point. It can be easily oxidized when exposed to air. C. Guo has reported the preparation of golden, gray, and black colors on aluminum surfaces [1]. However, the research on the formation of different colors just by adjusting the scanning speed is rare [2]. Here we modified the aluminum surfaces by scanning the femtosecond laser beam over the surfaces. The influence of the pulse overlap on processing result was studied in detail through observing the induced micro/nanostructure, and analyzing the composition of the modified surface. Results demonstrated that various structural colors can be generated by properly controlling the pulse overlap. Such an investigation is supposed to be able to provide a promising method for producing controllable structural colors.

Firstly, the possibility of producing structural colors on aluminum surfaces by altering the pulse overlap is examined. In experiments, the average laser fluence is 5.1 J/cm^2. The pulse overlap in the Y-direction is chosen to be -233% (hatch distance = 50 μm). The aluminum surfaces appear golden, white, gray and black when the overlap is set as -7, 33, 93, and 99%, respectively. The corresponding scan rate is 16, 10, 1, and 0.1 mm/s. The typical results are demonstrated in Fig. 3.1.

G. Li, *Bionic Functional Structures by Femtosecond Laser Micro/nanofabrication Technologies*, Springer Theses, https://doi.org/10.1007/978-981-13-0359-3_3

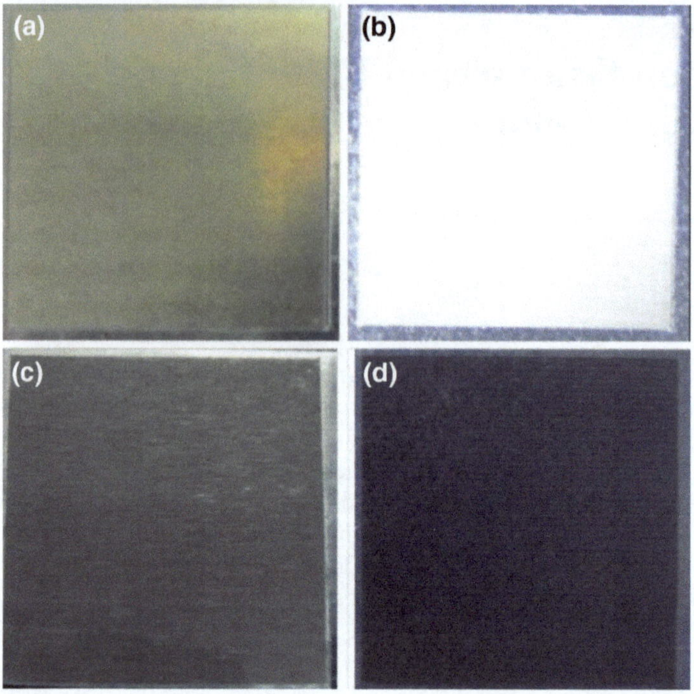

Fig. 3.1 Photograph of modified aluminum: **a** golden; **b** white; **c** gray; **d** black

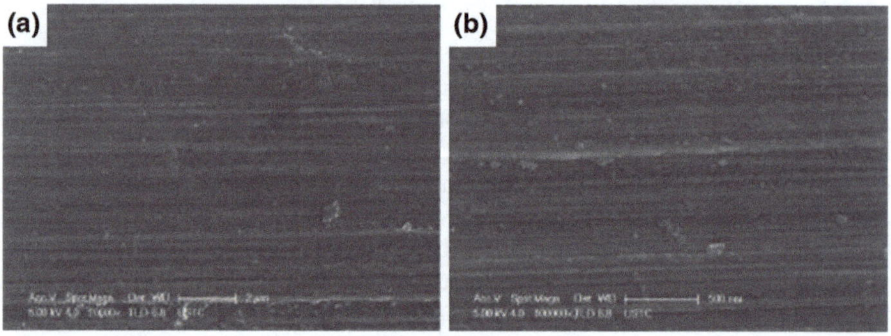

Fig. 3.2 SEM images of untreated aluminum surface: **a** 20,000 × magnification; **b** 100,000 × magnification

To identify how the surface structures are affected by the pulse overlaps, a detailed SEM study is performed on the aluminum sample surface. Figure 3.2 shows the SEM images of an unexposed aluminum surface. Figures 3.2, 3.3, 3.4 and 3.5 reveal the SEM images of irradiated targets at various pulse overlaps, which are −7, 33, 93 and 99%, respectively.

Fig. 3.3 SEM images of structures on golden aluminum surface

Fig. 3.4 SEM images of structures on white aluminum surface

It is clear from Fig. 3.3a that adopting the pulse overlap in x direction of −7% endows the modified aluminum surface with a smooth surface as compared to the untreated surface in Fig. 3.3. The more detailed structures can be seen from Fig. 3.3b–d. Many nanoparticles, whose diameters are in the order of hundreds of nanometers, are embedded on the substrate surface. As shown in Fig. 3.3c, these nanoparticles can be generally classified into two shapes [2]. One is like a flower

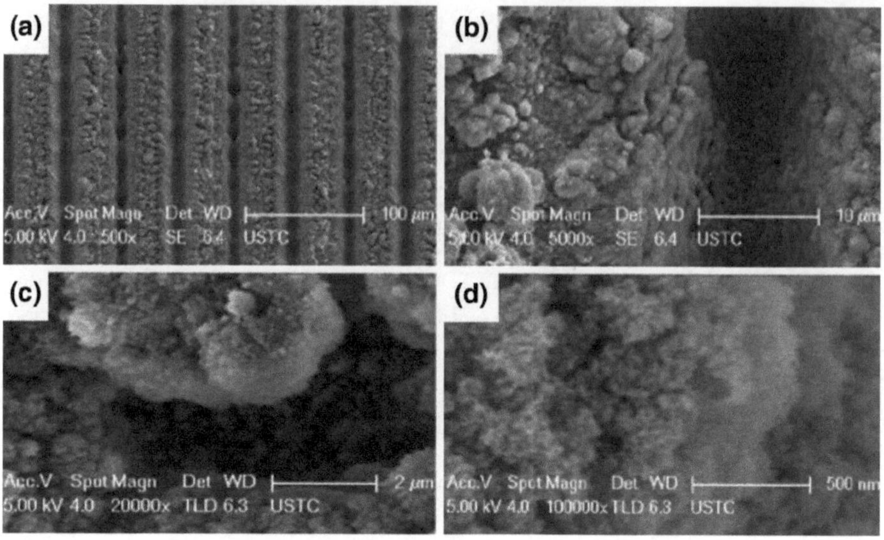

Fig. 3.5 SEM images of structures on gray aluminum surface

with diameter of approximately three hundred nanometers in diameter, and the other is like a sphere with diameter of approximately one hundred nanometers. Figure 3.3d confirms that these nanoparticles contain many protrusions and micro scale aggregates, which are formed by fusing nanoparticles together on the surface.

When the pulse overlap increases to 33%, slight stripe structures dispersed on the substrate surface appear as shown in Fig. 3.4a. Nonuniform cracks of different scale are also formed, as evidently displayed in Fig. 3.4b. When the surface morphology is magnified 20,000 times as shown in Fig. 3.4c, block-like structures and grainy structures can be found clearly. This is the evidence that the liquid phase exists in the laser ablation process. Higher pulse overlap makes more surface materials vaporize, therefore both the quantity and the size of spheres decrease and more fluffy structures are generated as shown in Fig. 3.4d.

Figure 3.5a represents the surface topography of representative groove structures generated by scanning a laser beam horizontally across a sample surface at a 93% overlap. The period of the groove is 50 μm, which is determined by the distance of adjacent horizontal scanning lines. The width of groove is about 15 μm, which is approximately equal to the diameter of the focused laser beam. Compared with the case of 33% overlap, the scale formed spherical and block-like nanoparticle aggregates much bigger as shown in Fig. 3.5b. More fluffy porous can be generated, which can be found in Fig. 3.5c and d.

When the overlap reaches 99%, the processing lines can produce deeper grooves on the aluminum surface as shown in Fig. 3.6a and b. From the cross-sectional SEM image, Fig. 3.6e, the depth of the grooves is measured to be ranging from 89 to 125 μm. Figure 3.6b–d demonstrate the black aluminum surface has a rich

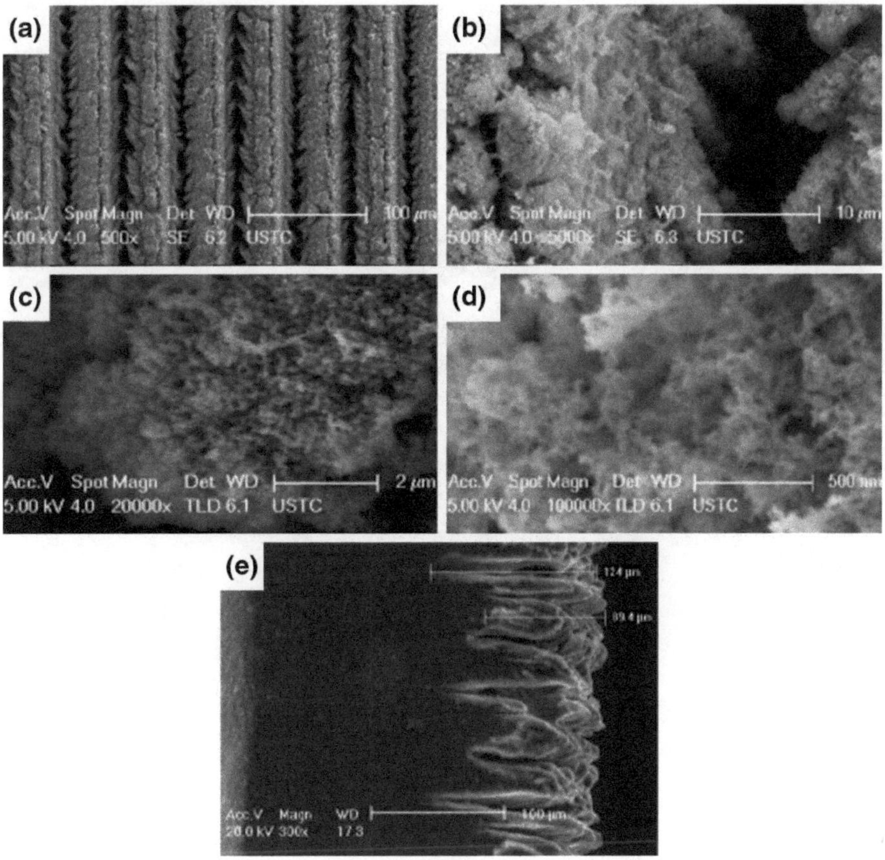

Fig. 3.6 The front and side SEM images of structures on black aluminum surface

variety of structures including types of nano- and micro-scale voids, nanoprotrusions, microscale aggregates formed by coalescence of nanoparticles [3].

In order to further study the influence of the pulse overlap on surface modifications, we also considered the case of the pulse overlap of 12% ($V = 13$ mm/s). Under such a condition, another kind of modified aluminum surfaces are fabricated, which exhibits different colors from different viewing angles as shown in Fig. 3.7. The SEM study shows that the entire surface of the aluminum is covered with femtosecond laser-induced periodic structures (LIPSS), and the spatial period of the LIPSS ranges from 500 to 550 nm. The LIPSSs can be considered as a kind of grating structures, which could play important roles in modifying the optical properties of metal surfaces in a very versatile way.

X-ray photoelectron spectroscopy (XPS) and X-ray diffraction spectra measurements are performed to investigate the chemical compositions and surface structures of irradiated aluminum sheets [3]. From the XPS spectra as shown in Fig. 3.8 and

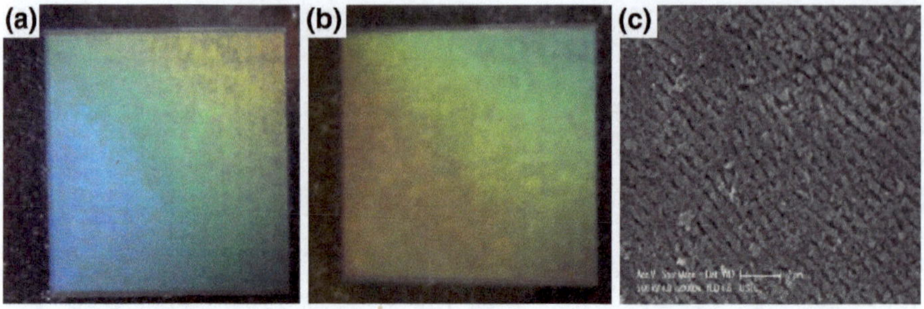

Fig. 3.7 Color aluminum formed at pulse overlap in x-direction of 12%: **a** viewed at 80°; **b** viewed at 60°; **c** SEM image of color aluminum

Fig. 3.8 XPS spectra for aluminum

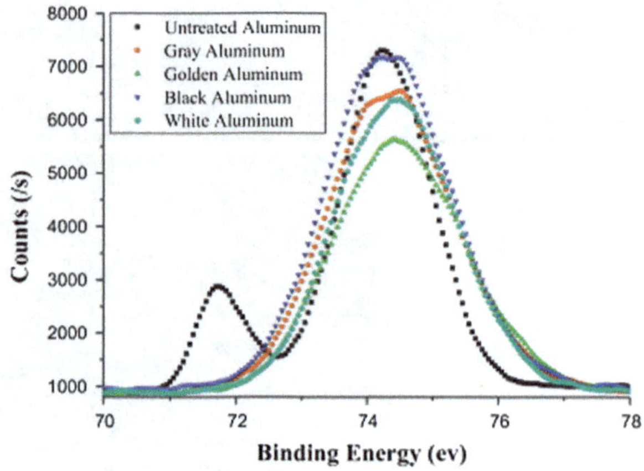

Table 3.1 XPS peaks for surface elements

	O1s (ev)	C1s (ev)	N1s (ev)	Al2p (ev)
Untreated aluminum	531.64	284.81	399.84	74.27
Golden aluminum	531.45	284.81	399.6	74.30
White aluminum	531.66	284.79	406.98	74.34
Gray aluminum	531.71	284.82	407.01	74.42
Black aluminum	531.77	284.79	407.18	74.45

peak table as shown in Table 3.1, it is found that for untreated aluminum, both the peaks at 71.7 eV BE and 74.27 eV BE can be detected, which are from the pure aluminum and the oxidized aluminum respectively. This result suggests the thickness of the natural oxide film should be less than 10 nm [3], for the maximum penetration depth of photoelectron is about 10 nm in XPS measurement. The peak at 74.27 eV is consistent with the Al2p binding energy.

Table 3.2 Surface element contents calculated from XPS survey spectra

	Oxygen (%)	Carbon (%)	Nitrogen (%)	Aluminum (%)
Untreated aluminum	40.38	43.64	1.45	14.54
Golden aluminum	49.92	32.19	1.67	15.22
White aluminum	55.52	27.21	1.33	15.95
Gray aluminum	57.62	24.21	1.79	16.38
Black aluminum	57.65	22.75	1.52	18.08

When the aluminum surfaces are irradiated with the femtosecond laser in air, only Al peak at 74.27 eV BE exists on the modified aluminum surfaces. Results indicate that the thickness of the oxidized film will increase more than 10 nm after the irradiation process.

All the samples not only have the expected O and Al elements, but also have a relatively high content of C1s and a small N1s. The contents of the Al and O increase with the increasing pulse overlap, which can be concluded from Table 3.2.

The surface structure of irradiated aluminum is identified by X ray diffraction spectra. The sharp and high intensity peaks from Al are shown in Fig. 3.9. In addition, Al_2O_3 and anorthic $Al(OH)_3$ are found on golden aluminum and white aluminum, while only Al_2O_3 can be found on the gray and black aluminum, which demonstrates that the oxidation reaction for aluminum is enhanced with the increase of pulse overlap [2, 3].

The XPS peaks table shows that all the C1s peaks are located at the same position (284.8 eV), which implicates that they have come from the same contribution. Carbon is a kind of extremely sensitive element therefore it is easy to detect by analyzing the XPS spectra. The origin of the carbon on the aluminum surfaces may come from the accumulated pollution during pre- and post-irradiation sample handling. In addition, we noted that the carbon content decreases significantly when the pulse overlap increases. This phenomenon can be explained from the surface topography of modified aluminums. The SEM images in Figs. 3.3, 3.4, 3.5 and 3.6 show that the aluminum sheets after laser irradiation are covered by particles, grooves and voids, which allow the carbon element to penetrate deeply into the aluminum surfaces and meanwhile the content on the surfaces decreases [2, 3].

As for the N1 s, we can see that its content in either the untreated aluminum surface or the modified ones is very weak, and there is no strong correlation between the N1s content and the pulse overlap. According to the N1s peak positions in Table 3.1, the origin of N can be attributed to two different contributions. The two peak positions (399.84 and 399.6 eV) are very close to the peak position (400 eV) for nitrogen in most organic materials, thus some part of N comes from organic matters. The other peak positions (406.98, 407.01, 407.18 eV) indicate that the other part of N might exists in the form of some nitrogen compound NO_3.

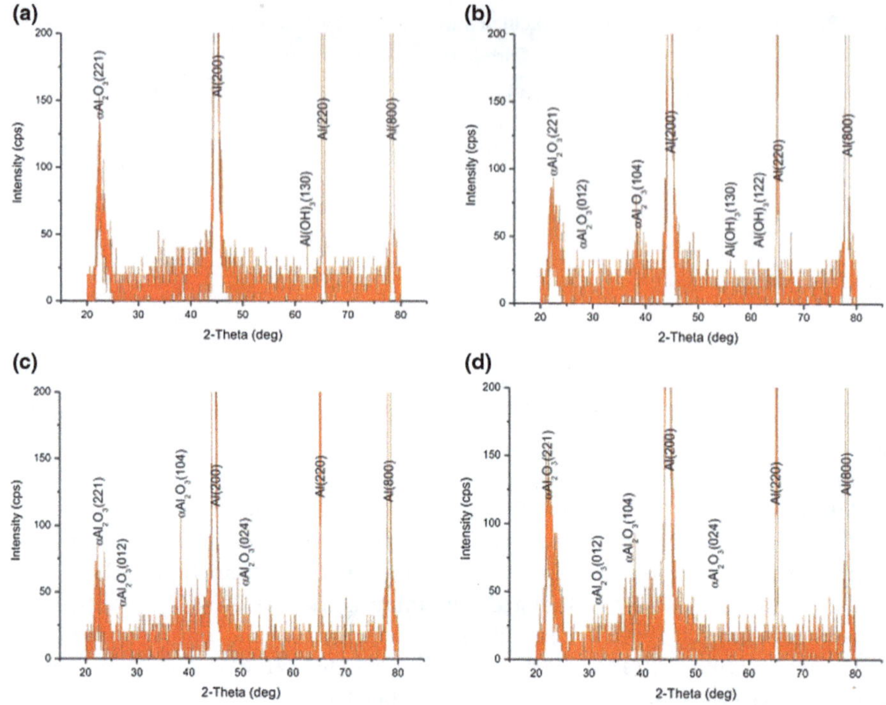

Fig. 3.9 XRD patterns of femtosecond laser treated aluminum surfaces: **a** golden aluminum; **b** white aluminum; **c** gray aluminum; **d** black aluminum

Therefore, colorizing aluminum surfaces can be mainly influenced by forming a variety of structures of different size, width/diameter, and/or depth. The surface structures, such as the microgratings, contribute for changing the optical properties of the aluminum surfaces, which result in different colors at different viewing angles because of the trapping of light inside the periodic microstructures and the angular dependence of Fresnel reflections. The random microstructures or nanostructures, play significant roles in forming colorful aluminum exhibit one kind of color in different angle of view [1, 2]. Many nanoparticles of hundreds of nanometer scale in diameter embedded on the substrate surface result in yellow color [2]. Block-like structures and grainy structures endow the aluminum surface with white color [2]. Grooves covered with types of nano- and microscale voids make the aluminum surface show gray or black color, and so on [2]. The phenomena of color change shows that the optical properties of the aluminum surface change with control of laser treatment parameters. Aluminum oxide appears pure white, which enhances the color of white, and weakens the other colors on modified aluminum surfaces [2].

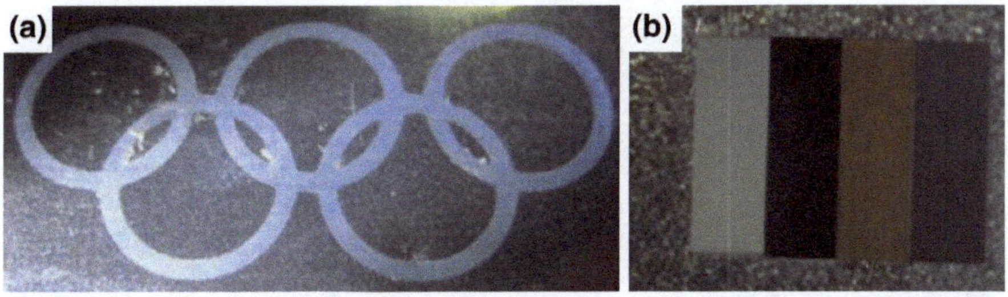

Fig. 3.10 **a** Olympic rings produced at the pulse overlap in x-direction of −27%, **b** combinatorial picture controllably produced

Based on above investigations, patterned colors on one surface can be flexibly produced by altering the pulse overlap. For verification, blue Olympic rings can be fabricated as shown in Fig. 3.10a by using an overlap of −27%. A modified aluminum sheet composed of white, black, golden and gray regions as shown in Fig. 3.10b can also be controllably produced by adopting the pulse overlaps as mentioned [2].

3.2 Diverse Displays for Multiple Color Patterns on Stainless Steel Surfaces

The creation of colorful metals produced by Laser Induced Periodic Surface Structures (LIPSS) has drawn considerable attention in recent years [4]. These LIPSSs can be considered as a kind of grating structures, which could play important roles in modifying the optical properties of metal surfaces in a very versatile way. The widely accepted theory believed that the formation of LIPSSs was the result of the interference between the incident laser beam and the surface-scattered wave. In addition, it is indicated that the ripples (LIPSSs) orientation usually is perpendicular to the polarization of laser pulse [4–6]. The correlation of ripples orientation and laser polarization offers us an opportunity to skillfully utilize the laser polarization as a control parameter to write diverse displayed patterns on metal surfaces. Specific color patterns can be generated due to the diffraction of light by producing polarization dependent structures. Color effects can be diversely displayed by altering the incidence angle of white light, which is used to irradiate the patterns constituted by ripples with different orientations [4–6] (Fig. 3.11). Although many works have been reported to explore how to use the laser polarization dependent ripples to obtain the needed optical diffractive effects, there are still few researches on systematical study of the combined influence of incident light angle and the ripples orientation on the color effects. Also there are few reports on the diverse displays for multi-patterns which are fabricated in the adjacent locations on the metal surface and constituted by ripples with different orientations. In this paper, we not only discussed the combined influence of incident light angle and the ripples orientation on the diversity of

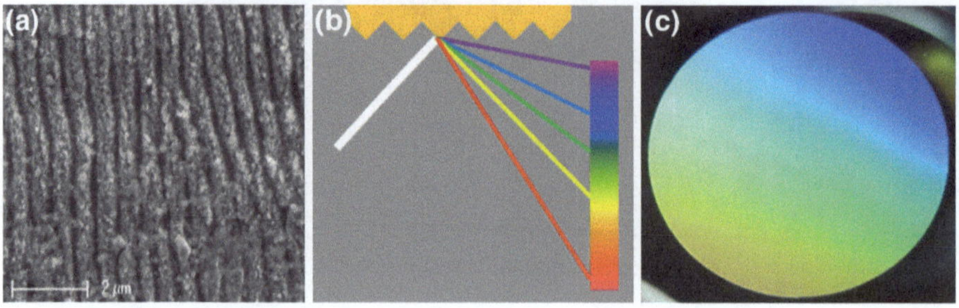

Fig. 3.11 **a** femtosecond laser induced periodic ripples on stainless steel surface; **b** schematic diagram of ripples diffraction; **c** photogragh of structural colors

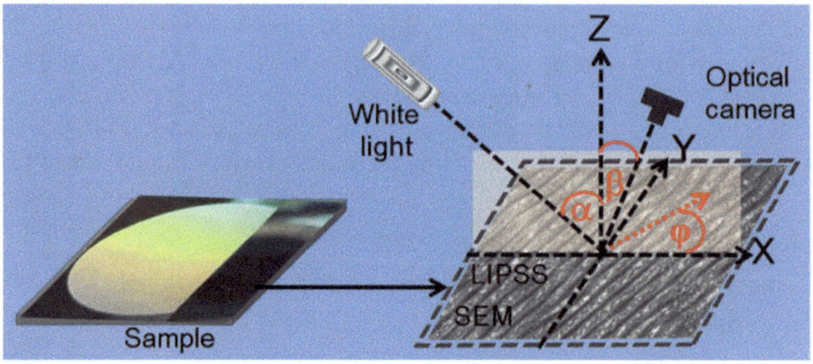

Fig. 3.12 Schematic illustration of color measuring system

structural colors, but also proposed the possibility of realizing the carry and diverse display of multi-patterns. Our investigation may find applications in the fields of information storage, identifying codes and anti-counterfeiting patterns and so on [4–6].

In this experiment, mirror polished 316 L stainless steel surface is used. Before the laser irradiation, the polarization direction of the laser pulse was adjusted by a linear Glan-Taylor polarizer. The scanning was realized by using a computer controlled high precision *XY* galvo mirror. The beam diameter focused onto the sample surface was about 20 μm. Schematic illustrations of measuring the optical properties were shown in Fig. 3.12. A digital optical camera (Cannon) was used to take the pictures of the structural colors. During the process of taking pictures, the surface ripples were irradiated by an unpolarized white light source from a LED lamp. For an objective evaluation of color effect, a spectrometer (Ocean Optics, USB2000) was used to measure the reflectance spectra diffracted by the formed ripples, if necessary. All measurements were carried out in dark to avoid stray lights from the surrounding environment.

Based on experimental investigation, a series of processing parameters were optimized to satisfy the optimum condition for generating ripple structures. In this experiment, the pulse overlap was set to 32%, the line space was fixed at 20 μm, and the laser fluence was chosen as 1.1 J/cm^2. Under the processing parameters, a circle pattern with diameter of 6 mm was fabricated on the stainless steel surface. The SEM images indicate that periodic ripples with period of $d = 540$ nm, and orientation to be perpendicular to the polarization orientation of laser pulse were formed.

In order to systematically investigate the influence of the incident angle of white light and the ripples orientation on the formation of structural colors, a color measuring system was configured as presented in Fig. 3.12. Figure 3.12 illustrates that the incident white light and the optical camera were placed in the *XZ* plane, with α and β orientation to the *Z*-direction on both sides, respectively. The sample constituted by the ripples with φ orientation to the *X*-direction was located in the *XY* plane. The range of α was from 0° to 90°, and that of φ was from 0° to 180°. The optical properties of the ripples can be considered to be close to those of reflective diffraction gratings. When white light is shined on the sample surface, diffraction gratings divide incident light into spectra with different wavelengths. The spectra are reflected differently by virtue of its wavelength-dependent refractive index. In this case, the lights with certain wavelengths are deflected from reaching the optical camera owing to the presence of structures. As a result, the reflected light with particular wavelength captured by the digital camera becomes sources of structural colors [4, 7, 8–11].

As the clarity of the taken colors depend on the observation angle to some extent, many measurements were performed to find the optimal location of the digital camera. It was found that the shadow of the camera could be seen in the sample surface when the viewing angle β was less than approximately 20° for the specular reflection of the mirror polished stainless steel surface. On the other hand, if the viewing angle was larger than approximately 60°, part of the taken colors was dim, which was unsatisfactory. Hence, observation of the clearest color on the basis of analyzing the optical images was adopted as a criterion for the optimal viewing angle. On this basis, an angle of about 25° was chosen, and this was kept constant for all measurements [4].

Under different incident light angle α and ripples orientation angle φ, the pictures of the diverse colors were taken. The pictures were cropped and arranged in order, as shown in Fig. 3.13. It is revealed that the colors are strongly dependent on the incident light angle α and ripples orientation angle φ. In addition, it is noticed that the colors mainly appear in the central zone, with "*V*" shaped pro-file, nearly spanning the whole visible spectrum. On the contrary, in other regions, the colors disappear, or display completely dark.

To explain this phenomenon, the ripples are regarded as rectangular reflection grating, and the theoretical analysis based on the grating diffraction is performed. The diffracted light wavelengths λ that match the different colors can be calculated with the following diffraction Eq. 3.1 [5, 7]:

$$n\lambda = d(\sin\alpha \cdot \sin\varphi + \sin\beta) \tag{3.1}$$

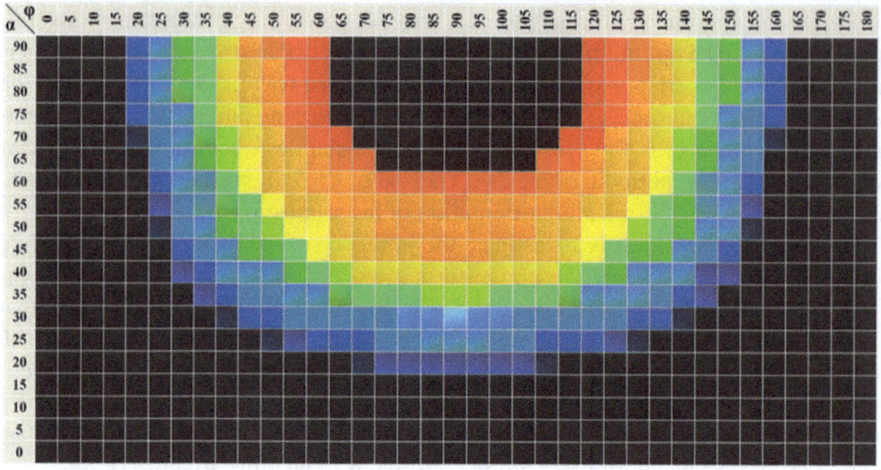

Fig. 3.13 Photograph of the structural colors observed at different incidence angles and ripples orientations

Fig. 3.14 Schematic diagram of grating diffraction

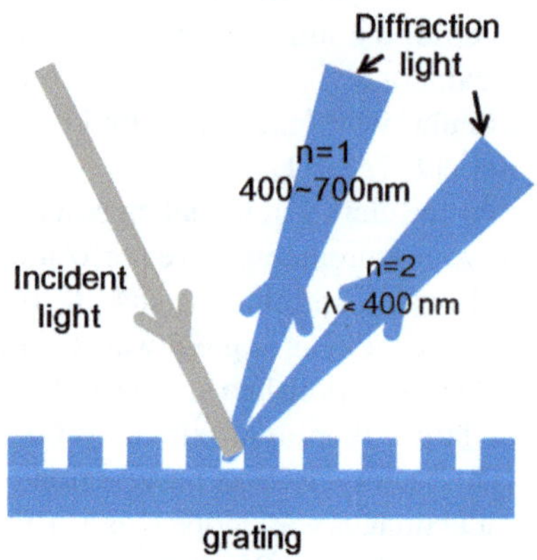

where n is the order of diffraction, which is an integer. λ is the wavelength of white light wavelength, and the range is from approximately 400–700 nm, to cover the visible color spectrum. d indicates the period of the grating (540 nm). The range of the $d(\sin \alpha \cdot \sin \varphi + \sin \beta)$ in our case is from 228 nm to 768 nm, hence only the first-order diffraction from the grating can be selected, namely $n = 1$ (Fig. 3.14). Then the calculated wavelengths are transformed into the corresponding colors, and the result is presented in Fig. 3.15. Obviously, for each specific incident angle of white light, diverse colors can be viewed at different angles.

In detail, we take the incident angle of 90° for example. In this case, the diffraction equation can be rewritten as

$$\lambda = 540 \sin \varphi + 228 \quad (3.2)$$

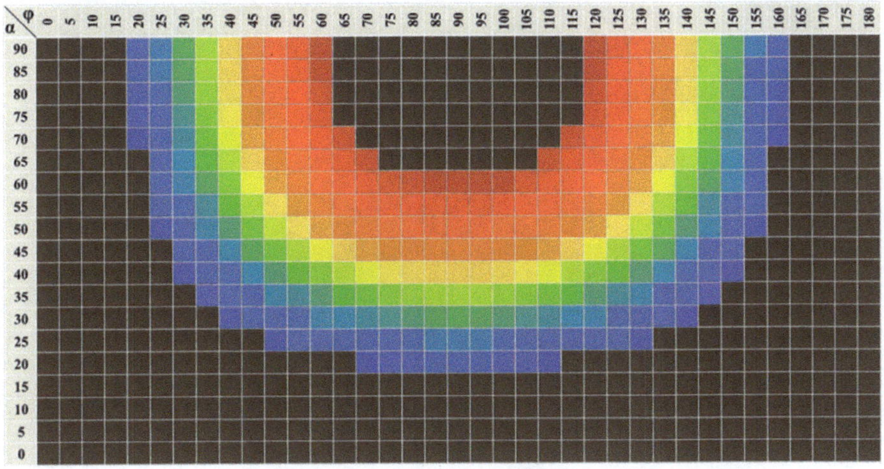

Fig. 3.15 The simulation results of the structural colors

The diffraction condition is $0.3 \leq \sin\varphi \leq 0.874$. Therefore, the corresponding range of φ is from 17.46° to 60.93°, and from 119.07° to 162.54°. For other orientation angle ranges, there is no visible wavelength found to fulfill the diffraction condition, hence no structural color can be observed.

A systematic comparison between the experimental results and the simulation ones reveals that some color differences can be found. This can be attributed to the following reasons: (1) even though the processing parameters were carefully chosen, LIPSS itself is not a very uniform periodic structure which has its intrinsic limitation in terms of grating uniformity. This nonuniformity can causes the differences in the light modulation; (2) the irregular nanoparticles formed on the top of the LIPSSs can absorb or reflect certain wavelengths of light; (3) the discordance between environmental conditions and the measurement system; (4) the fluctuation and distortion of light source. Although the experimental results are not very perfect compared to the model, this color reported in this work demonstrate angle sensitive.

From the above discussion, it is evident that the colors can be periodically displayed if ripples orientation angle φ is ranged from $-\infty$ to $+\infty$. Moreover, in this case, all the above-mentioned colors in Fig. 3.13 can appear at different locations if another viewing angle β is chosen.

The following section provides an example to illustrate a possible application of this finding, which is done by introducing the diverse display of multipattern. For this purpose, four numbered circle patterns, arranged in two rows and two columns, were fabricated on the stainless steel surface, as shown in Fig. 3.16a. The diameter of each pattern is 6 mm. The surface structures are constituted by ripples with orientations of 0°, 30°, 60°, 90°, respectively, which are designed by controlling the laser polarization. The polarization angles for samples 1–4 are 90°, 60°, 30°, and 0°, respectively, as shown in the SEM images of Fig. 3.16b. In the initial position, the ripples orientation of the No. 1 pattern is alignment with the X-direction.

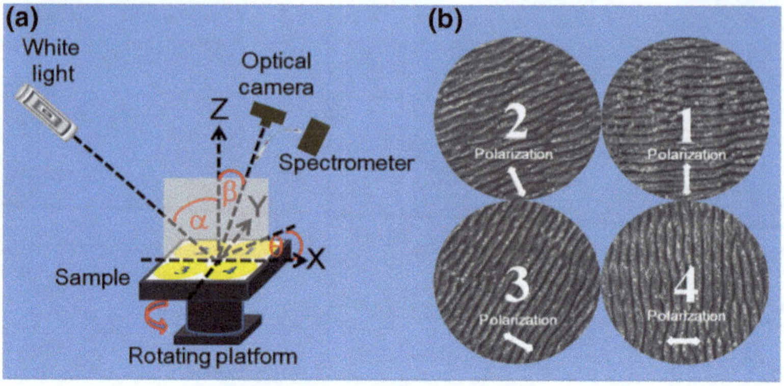

Fig. 3.16 **a** Schematic illustration of measuring the optical property; **b** SEM images of the ripples with orientations of 90°, 30°, 60°, 90°, respectively

The change of the rotating sample angle θ can be realized by rotating the platform anticlockwise around the Z-direction. In this way, ripples orientations of all these patterns can be varied. Correspondingly, the diffraction equation for each pattern can be written in sequence as

$$n_1\lambda_1 = d(\sin\alpha \cdot \sin\theta + \sin\beta) \tag{3.3}$$

$$n_2\lambda_2 = d(\sin\alpha \cdot \sin(\theta + 30°) + \sin\beta) \tag{3.4}$$

$$n_3\lambda_3 = d(\sin\alpha \cdot \sin(\theta + 60°) + \sin\beta) \tag{3.5}$$

$$n_4\lambda_4 = d(\sin\alpha \cdot \sin(\theta + 90°) + \sin\beta) \tag{3.6}$$

It should be noted that enhanced colors can be observed on the pattern when the corresponding diffraction equations meet condition $400 \leq n\lambda \leq 700$ nm, ($n = 1$). In this way, one or more patterns can be distinguished selectively from all the patterns by the corresponding colors. It is desirable to use a computer to perfectly analyze this selectivity. In this analysis, four cases of "one out of four pattern", six "two out of four patterns", four "three out of four patterns", a "none" and an "all" can be obtained.

For verification, the related experiments are conducted by studying the structural colors of the patterns. The home-made setup is illustrated in Fig. 3.16a. By conducting the experiment carefully, it is found that the desired patterns can be observed with enhanced colors by adjusting the incident light angle and the rotating sample angle, as shown in Fig. 3.17. Meanwhile, the rest cannot be observed for the colors disappeared.

In addition to the cases shown in Fig. 3.17, the sample can present the same patterns with different enhanced colors at other incident light angle and rotating sample angle. A typical experiment shown in Fig. 3.18 analyzes the impact of the angle sensitivity on the color effects for the same patterns. In Fig. 3.18, the No. 1 pattern

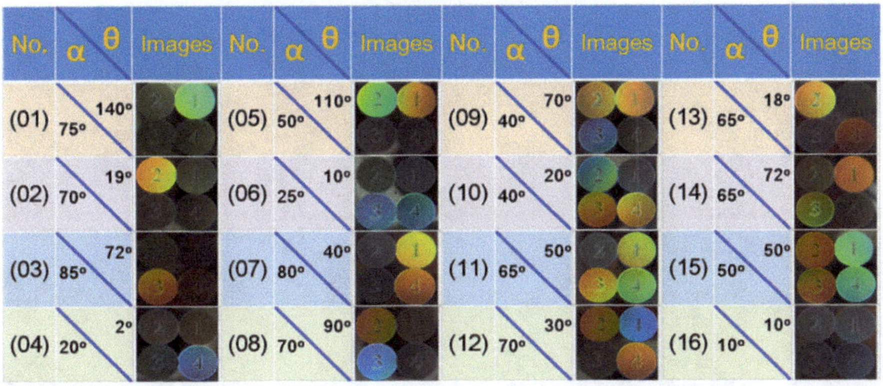

Fig. 3.17 Multi-patterns displayed with different structural colors

Fig. 3.18 Images and spectrum of No. 1 pattern measured at different angles

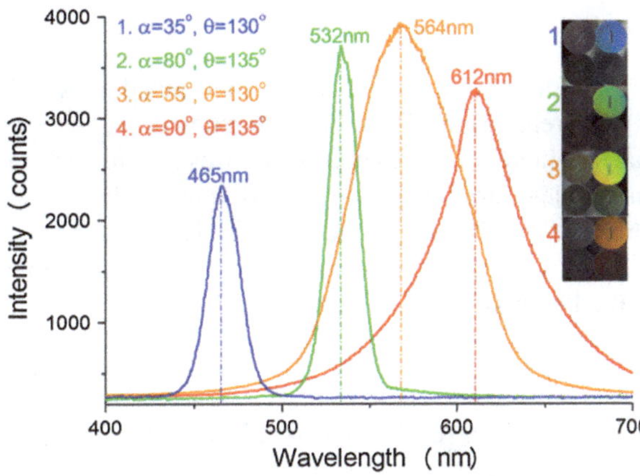

exhibits with blue, green, yellow and orange at the given angles. Since the main reason for the colors is the diffracted light received from the observed pattern by camera, the spectral properties for the pattern are investigated with a spectral measurement system (Fig. 3.16a). The reflectance spectrum of the pattern is made as a function of the visible wavelengths, as shown in Fig. 3.18. The measurements give the spectral information for the measured color. This experiment reveals that the desired patterns can be displayed with different enhanced colors, demonstrating strong angle sensitivity. Assuming that enhanced colors represent different information, multiple types of information stored with the mentioned method can only be correctly identified in certain condition. Such an investigation can provide a promising method to fabricate patterns containing security information with various colors which are not only readable but also difficult to be imitated. Hence, the technique may find applications in meeting the urgent requirements for forgery protection, information storage, and identifying codes and so on.

Fig. 3.19 The application of structural color in artwares

In addition to the applications mentioned above, this technique can also be used in the preparation of artwares. As shown in Fig. 3.19, "Eight Trigrams" and "Zodiac", which are representative figures in Chinese traditional culture, are inscribed on stainless steel surface with different ripples directions. Under the irradiation of illumination, the pattern can be selectly displayed with different colors. Combining with the advanced computer control technology, more complex color patterns, such as the butterfly (Fig. 3.19), can also be fabricated.

3.3 The Preparation of Three-Dimensional (3D) Patterns

In this section, we propose the preparation of three-dimensional (3D) patterns with femtosend laser coloring technology. Because of the opacity of the metal materials, it is impossible to prepare patterns at different layers by focusing the laser beam to different depths [4, 8]. The key to prepare three-dimensional (3D) patterns on metal surface is fixing the positions of the multiple patterns. For example, Chinese characters "中国" in Fig. 3.20a is visually two-dimensional. Amazedly, the two-dimensional pattern can be changed to three-dimensional one if the same pattern is overlayed and then translated repeatedly along the XY directions at certain scale, as shown in Fig. 3.20b.

Based on the above discussion, the three-dimensional color patterns can be produced on stainless steel surface with femtosecond laser coloring technology (Fig. 3.21). The advantage of this technology lies in the preparing the three-dimensional and complicated patterns on metal surfaces flexibly, which has wide applications in the fields of the industrial arts.

Fig. 3.20 Schematic diagram of designing three-dimensional patterns

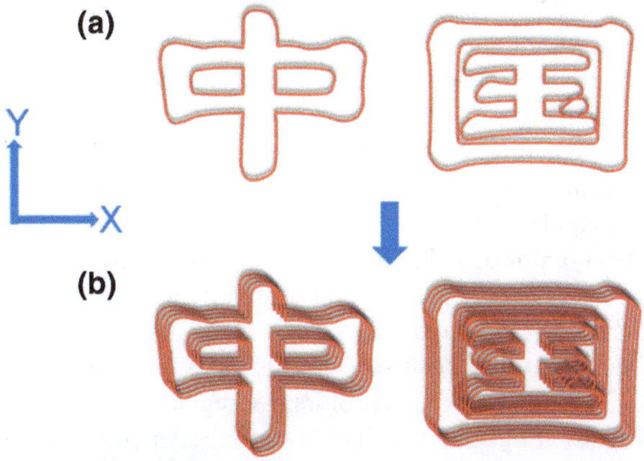

Fig. 3.21 The preparation of three-dimensional color patterns with femtosecond laser on stainless steel surface

3.4 Coloring Metal Surface with Different Laser Wavelengths

We find that the ripple's period is directly dependent on the laser wavelength and the materials. To the best of our knowledge, most papers indicate that the colors are produced with laser wavelength of about 800 nm, and the corresponding ripples periods ranged from 480 to 580 nm. However, the structural colors formed by other laser wavelength are less researched. In this section, a systematic study is performed on the dependence of structural colors on the laser wavelengths which ranged from 400 to 2200 nm. It is found that completely different colors can be obtained by different types of ripples induced with varying laser wavelengths. The discussion in this paper suggests femtosecond laser wavelength as a potential parameter to fabricate various colors and ultimately leads to controlling the optical properties of the metals.

The experiments were conducted to study the colors which are formed by a variety of subwavelength ripples induced with varying laser wavelengths. The laser wavelength, a key processing parameter, is modulated by the optical parametric amplifier (OPA, TOPAS-800-fs), and the modulation range is from 240 to 2600 nm.

It indicates that ripples with various kinds of periods can be induced by different laser wavelength under suitable laser fluence and scanning speed. The exact calculation for spatial period Λ is still difficult. However, the goal of our work was to obtain the needed ripple periods by precisely controlling the laser wavelengths experimentally. It is found that uniform ripples with different spatial periods can be obtained at optimized laser fluence in the range of 0.96–1.02 J/cm^2, scanning speed of 13 mm/s. The ripple spatial periods (Λ) as a function of the incident laser wavelength (λ_L) are illustrated in Fig. 3.22a. The partial SEM images of the ripples formed on stainless steel surface are presented in Fig. 3.22b. The results show that the spatial periods of LIPSS depend on the laser fluence and the incident laser wavelength. The formation mechanism has been studied by Sakabe et al. and Okamuro et al.

To study the formed colors, we regard the ripples as gratings. It is known that the diffracted light by the ripples strongly depend on the light source directions and the ripple directions. It indicates that the horizontally mounted sample, rotated around Z-direction, is illuminated by an unpolarized light source which is placed in the XZ plane, with α orientation to the Z-direction, equivalent to the day-light spectrum. φ is the ripple direction to X-direction. Both the ranges of α and φ are from 0° to 90°. Then, the diffracted light is observed with a β orientation to the Z-direction, at the contrary direction of light source. In our study, the viewing angle is fixed at 20°.

The gratings with different spatial periods having a size of 5 mm × 5 mm are examined under the color measuring system. The results are shown in Fig. 3.23. It is indicated that the colors are different at different coordinates for a given ripple period. In addition, the observed colors are also different at the same coordinates for different ripple periods.

Based on the color measuring system, diffraction Eq. 3.1, and schematic diagram of grating diffraction (Fig. 3.14), a related theoretical simulation can be performed through the analysis of the diffraction equations with different periods. The wavelength of second-order diffraction light is <400 nm which cannot be observed for most spatial periods. In addition, the intensity of the second-order diffraction light or even higher orders is so weak that it can be ignored. Hence, only the first-order diffraction from the grating can be selected, namely $n = 1$. From the calculated spatial periods matching the different diffracted wavelengths, it is observed that the processed areas can exhibit structural colors when the period Λ is >298 and <2047 nm. When the ripples period is beyond this scope, there is no wavelength λ found to fulfill the diffraction conditions, thus no color can be observed. In addition, the colors can cover the whole visible spectrum if the spatial period is >522 and <1170 nm. As described above, we can divide the periods into three scopes, namely 298 to 522 nm, 522 to 1170 nm and 1170 to 2047 nm, to separately study the influence of periods on the color effects.

Firstly, the spatial periods ranged from 298 to 522 nm is discussed. By solving the diffraction equation, α and φ matching different periods are calculated. Then, the calculated α and φ are set to horizontal and vertical coordinates, respectively. The segmental results are presented in Fig. 3.24. One can see that the silhouette which is besieged by α and φ meeting the diffraction equation (marked blue in figures,

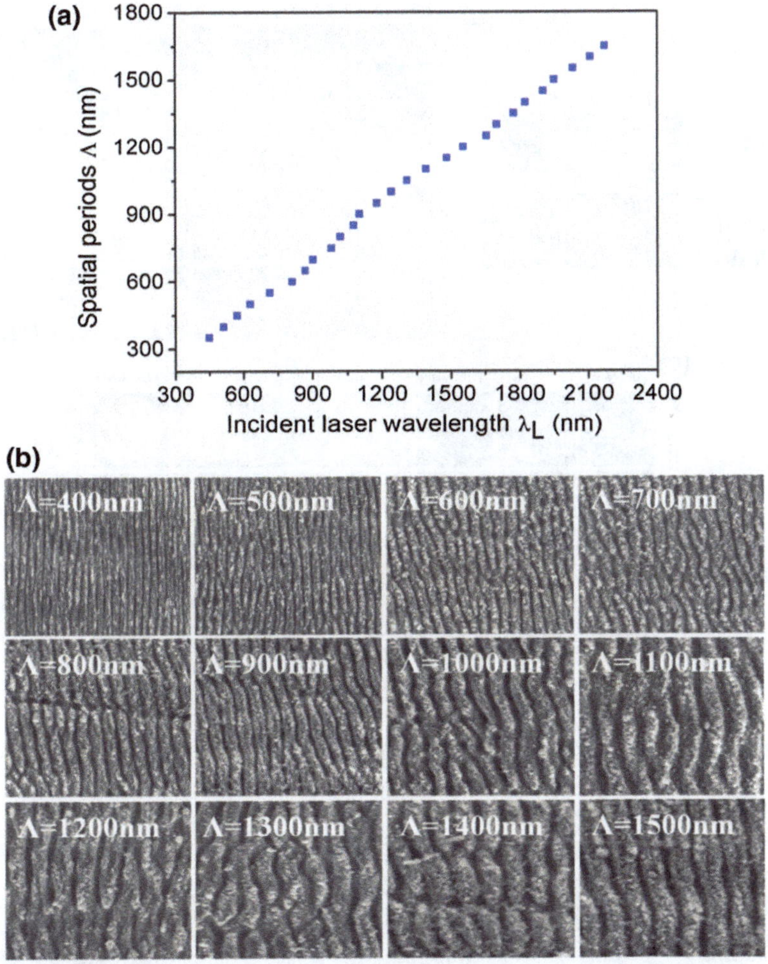

Fig. 3.22 **a** Ripple spatial periods as a function of the incident laser wavelength. **b** Partial SEM images for the laser-induced periodic surface structures by different laser wavelengths

hereinafter referred to as "silhouette") present a completely symmetric about the angular bisector of the coordinate system.

By comparison, obviously, at low spatial period of 300 nm, the diffracted light can cover only a few nanometers, namely from $\lambda = 400$ to 403 nm, as shown in Fig. 3.24a. As the periods increase, a broader spectrum appears and the maximum value of the spectrum increases (Fig. 3.24b–c). When the periods are increased to 521.6 nm, the whole visible spectrum can be covered (Fig. 3.24c). Moreover, the locations of the silhouettes are getting closer to coordinate axes.

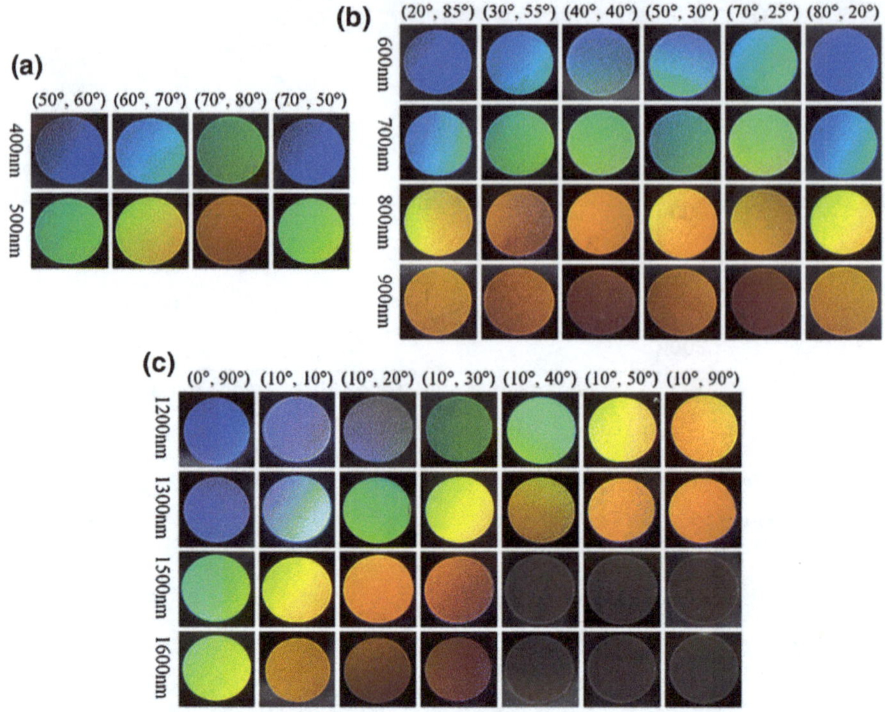

Fig. 3.23 Photographs of the structural colors formed with different spatial periods

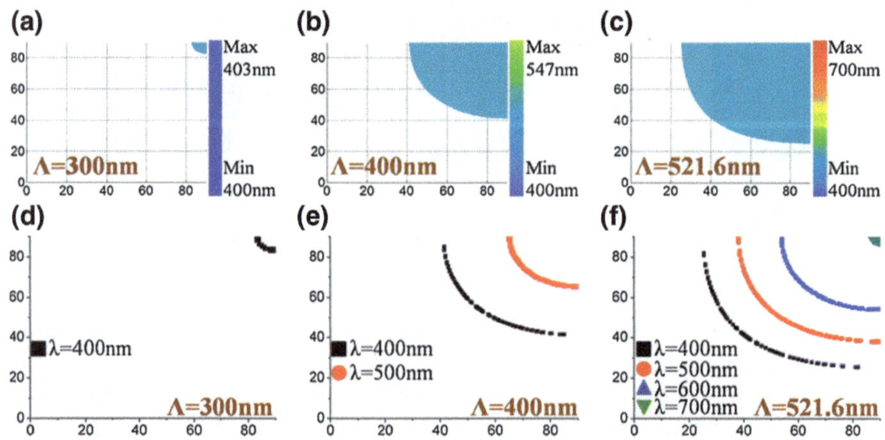

Fig. 3.24 Calculated coordinate meeting (a, b, c) and spectrum (d, e, f) for the structural colors formed under the spatial periods of 300 and 522 nm. The bars on the right of **a**, **b** and **c** indicate the spectrum range

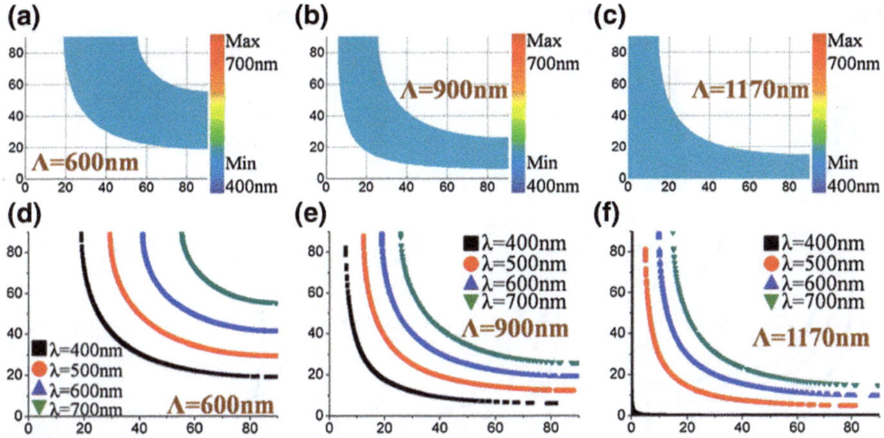

Fig. 3.25 Calculated coordinate and spectrum for structural colors formed under spatial periods of 400 and 500 nm

In order to get a deep insight into the influence of the spatial periods on the transfer of the spectrum, we calculate the locations of spectrum λ = 400, 500, 600, and 700 nm in α, φ coordinate system, which are presented in Fig. 3.24d–f. For Λ = 300 nm, it indicates that only the spectrum of λ = 400 nm can be observed near (90°, 90°), presenting an arc-shaped line, which is symmetric about the angular bisector of the coordinate system. When Λ is increased to 400 nm, the coordinate areas for the spectrum λ = 400 nm becomes broader. In addition, the spectrum λ = 500 nm appears. However, it is noticed that the two spectra are away from each other, which is demonstrated in Fig. 3.24e. As Λ becomes 521.6 nm, it is the lower limit of the Λ meeting the condition to cover the whole visible spectrum. In this case, the four spectra lines appear together. In contrast, the spectra lines are located very close to each other, and they have the trend to get close to the origin of coordinates.

Secondly, the influence of spatial periods ranged from 522 to 1170 nm on the diffracted spectrum is discussed, and the results are shown in Fig. 3.25. In this period scope, it is found that no matter what the location of the silhouette is, diffracted light can cover the whole spectrum (Fig. 3.25a–c). Similar to the previous, the locations of λ = 400, 500, 600, and 700 nm are getting closer to the origin of coordinates, and the spectral lines are closer to each other (Fig. 3.25d–e).

In addition to the two period scopes discussed above, it is indicated that the spatial periods ranged from 1170 to 2047 nm also affect the diffracted spectrum greatly. The silhouette is becoming narrower with the increasing periods, leading to the diffracted light only covers a part of the visible spectrum, and the lower limit of the spectrum rises and the red shifts occur. The partial results are graphically expressed in Fig. 3.26.

As described above, the coordinate areas and the spectral ranges of the colors are affected by the spatial periods of the ripples. In order to demonstrate this effect, we have calculated the ratio of silhouette area to total coordinate area and the extremum

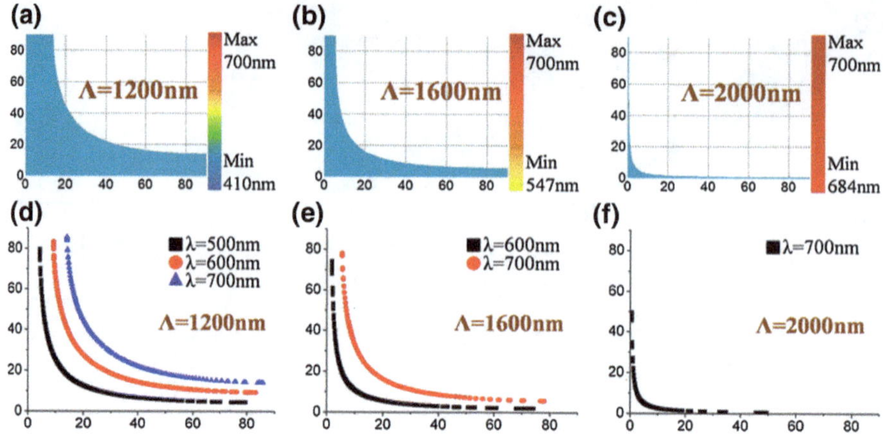

Fig. 3.26 Calculated coordinate and spectrum for the structural colors formed under the spatial periods ranged from 1200 to 2000 nm. The structural colors produced under spatial periods of 1200, 1300, 1500 and 1600 nm

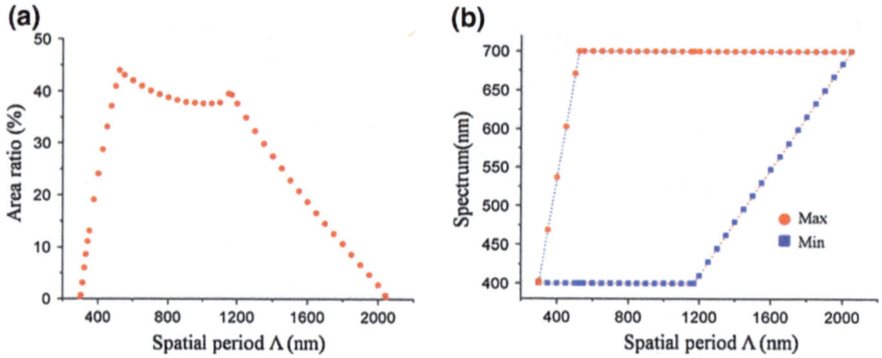

Fig. 3.27 **a** Ratio of silhouette area to total coordinate area; **b** Maximum value and the minimum value of the spectrum of the structural colors formed with different spatial periods

of the spectrum formed under the whole spatial periods range. The results are illustrated in Fig. 3.27. From Fig. 3.27a, it is worth noticed that the ratio has two maximum values of 43.9 and 39.5%, which are obtained at spatial periods of 522 and 1170 nm, respectively. As for the spectrum range of the diffracted light, it can be found from Fig. 3.27b that the spectrum range has a linear relation to the spatial periods of ripples.

Based on the above-mentioned discussion, it is not difficult to understand why the surfaces composed of ripples with different spatial periods exhibit very different colors. Furthermore, from Fig. 3.23, it is found that the colors are gradually changed from blue to red with the increase of periods under the same ripple orientation angle and incident light angle. At the approximate symmetric coordinate points, such as the points of (50°, 60°) and (70°, 50°) in Fig. 3.26a (20°, 85°) and (80°, 20°) in Fig. 3.23b, the colors are nearly identical. This can be attributed to the symmetry of the spectral lines about the angle bisector for coordinate. In Fig. 3.26c, it is obviously observed that the colors formed by the ripples with spatial period of 1200 and 1300 nm can cover most of the visible spectrum. While for the colors formed under spatial periods of 1500, and 1600 nm, only the spectrum longer wavelength can be covered. In addition, at many points, such as (10°, 40°), (10°, 50°), (10°, 90°) and so on, there is no wavelength of white light found to fulfill the diffraction conditions, thus no color can be observed. As for the ripples with spatial periods longer than 1600 nm, the formed colors are not experimentally discussed, because the power of the laser at longer incident wavelength is reduced too rapidly to induce uniform ripples. However, the results give the experimental evidence that color effects depend on the ripple period strongly.

As described above, the theoretical and experimental analyses demonstrate that the colors are affected greatly by the ripples with different spatial periods. The study proposes the possibility of marking the metal surfaces with various colors which are formed by different laser wavelengths.

References

1. Vorobyev AY, Guo C. Colorizing metals with femtosecond laser pulses. Appl Phys Lett. 2008;92(4):041914.
2. Li G, Li J, Yang L, et al. Evolution of aluminum surface irradiated by femtosecond laser pulses with different pulse overlaps. Appl Surf Sci. 2013;276:203–9.
3. Li X, Yuan C, Yang H, et al. Morphology and composition on Al surface irradiated by femtosecond laser pulses. Appl Surf Sci. 2010;256(13):4344–9.
4. Li G, Li J, Hu Y, et al. Realization of diverse displays for multiple color patterns on metal surfaces. Appl Surf Sci. 2014;316:451–5.
5. Dusser B, Sagan Z, Soder H, et al. Controlled nanostructrures formation by ultra fast laser pulses for color marking. Opt Express. 2010;18(3):2913–24.
6. Yao J, Zhang C, Liu H, et al. Selective appearance of several laser-induced periodic surface structure patterns on a metal surface using structural colors produced by femtosecond laser pulses. Appl Surf Sci. 2012;258(19):7625–32.
7. Li J, Li G, Hu Y, et al. Selective display of multiple patterns encoded with different oriented ripples using femtosecond laser. Opt Laser Technol. 2015;71:85–8.
8. Li G, Li J, Hu Y, et al. Femtosecond laser color marking stainless steel surface with different wavelengths. Appl Phys A. 2015;118(4):1189–96.
9. Huang J, Wang X, Wang ZL. Controlled replication of butterfly wings for achieving tunable photonic properties. Nano Lett. 2006;6(10):2325–31.
10. Ganbavle VV, Bangi UKH, Latthe SS, et al. Self-cleaning silica coatings on glass by single step sol-gel route. Surf Coat Technol. 2011;205(23):5338–44.
11. Woodward I, Schofield WCE, Roucoules V, et al. Super-hydrophobic surfaces produced by plasma fluorination of polybutadiene films. Langmuir. 2003;19(8):3432–8.

Chapter 4
Facile Fabrication of Functional PDMS Surfaces via Femtosecond Laser Textured Templating

Laser structuring technique has been intensively studied because of its advantageous properties of non-polluting, low-cost and high materials utilization. Lasers can play a role of an alchemist on coloring metals with diverse colors, as reported. Both individual and iridescence colors can be obtained by lasers on metals surfaces. Laser induced periodic surface structures (LIPSS) are regarded as the main cause of the color. Furthermore, surfaces with different wetting properties can be achieved by lasers in metals or polymers. Roughness factor is the origin of the wetting behavior. However, the major disadvantage of the laser structuring is its low efficiency caused by direct-writing technique, in which a raster-scanning strategy is commonly applied. High-throughput production of functional surfaces is intensely desirable and remains as an ultimate challenge. In this work, we functionalize the metal surfaces with different colors by adopting femtosecond laser texturing technique, and the treated metals are subsequently used as the master molds for efficient production of hydrophobic surfaces of PDMS with an extremely strong adhesion force. During the femtosecond laser processing, the laser pulse overlap can be easily and precisely controlled. We found that the laser pulse overlap not only affects the color appearance of the metal templates, but also changes the contact angle (CA) of the polymer surfaces in a relatively wide range. The water adhesion properties of the replicated polymer surfaces are demonstrated. Aluminum is chosen to be functionalized here as it is widely utilized in product packages, machine shells and decorative material. Firstly, the aluminum surfaces are treated by femtosecond laser scanning. Secondly, a soft lithography process is carried out to transfer the complementary micro/nanostructures onto polydimethylsiloxane (PDMS) surfaces under ambient conditions [1].

4.1 Optical and Morphology Properties of Aluminum Templates

Figure 4.1a, b are golden and black aluminum surfaces realized by the extremely low (−10%) and high (99.5%) pulse overlap, corresponding to the scanning velocity of 22 and 0.1 mm/s [1]. When varying the pulse overlap to 35 and 95%, the treated

G. Li, *Bionic Functional Structures by Femtosecond Laser Micro/nanofabrication Technologies*, Springer Theses, https://doi.org/10.1007/978-981-13-0359-3_4

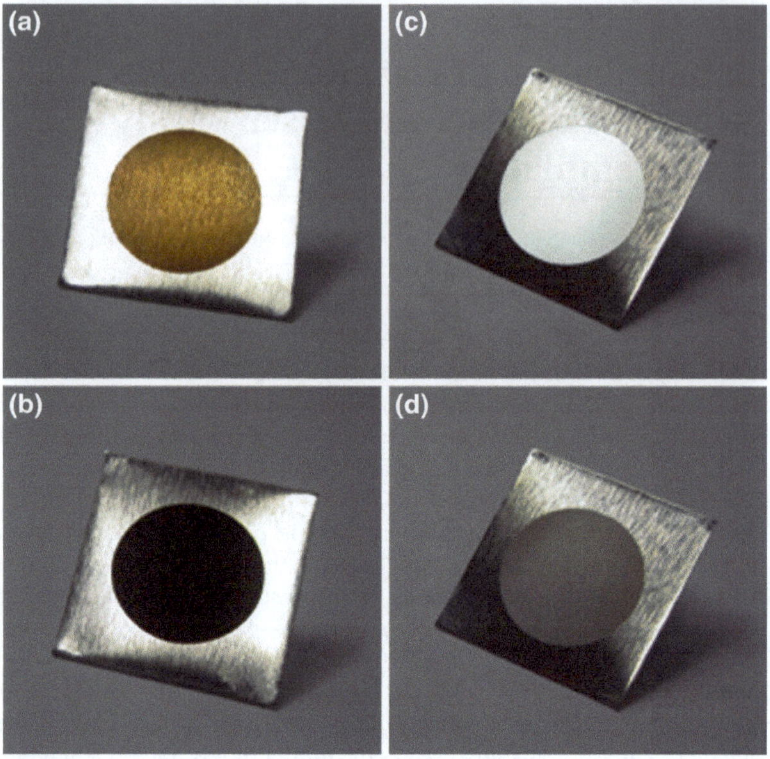

Fig. 4.1 **a–d** Photographs of laser treated aluminum pieces with golden, black, white and gray colors, respectively

aluminum surfaces appear to be pure white (Fig. 4.1c) and gray (Fig. 4.1d). The average laser fluence is set to be 5.1 J/cm^2. The scanning hatch in the longitudinal direction is 50 μm. Different from the iridescent structural colors obtained by other approaches, all the surfaces exhibit the same color under any view angles. The results are consistent with that has been reported before.

Figure 4.2 is the surface morphology of the corresponding golden, white, gray and black aluminum, examined by a scanning electron microscope [1] (SEM). The insets of each image show the magnified details. From the SEM study, we can see that the mesoscale period remains unchanged at 50 μm, which is corresponding to the Y-direction translation step. For the golden aluminum, some smooth scaly aggregates appear along the center of the laser beam scanning path, at the side of which a lot of particles with the size scale of hundreds nanometers are induced by the relatively low laser irradiation dose. The smooth aggregates reveal that liquid phase exists during the laser processing and then solidifies due to the low melting point of aluminum. Groove structures appear on the white aluminum surface, produced by the laser scanning over the surface. The groove width is around 20 μm, consistent with the focus spot size. Besides the nanoparticles identical to the golden aluminum, some fluffy structures

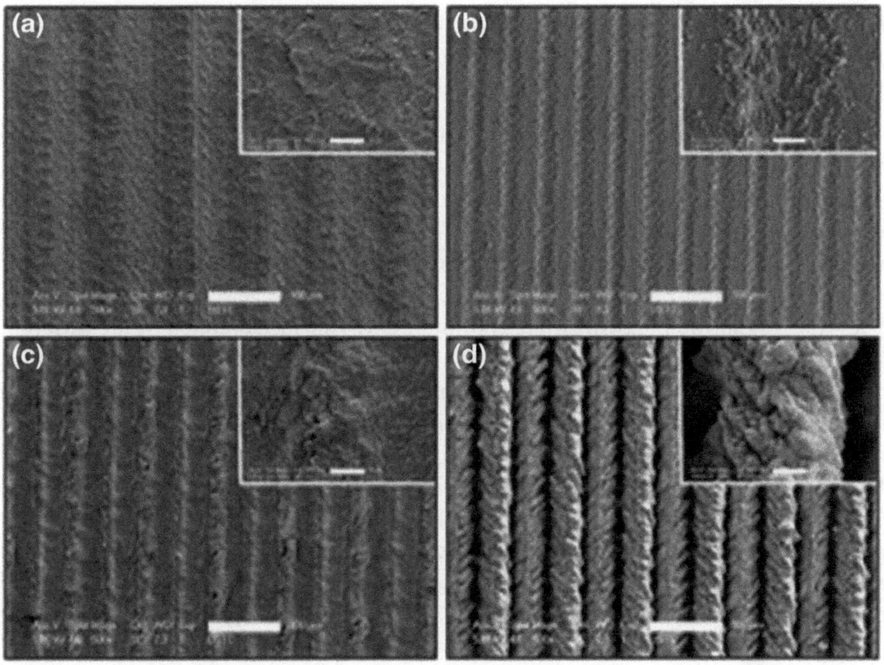

Fig. 4.2 SEM images of **a** golden, **b** white, **c** gray and **d** black aluminums. Scale bars: 100 μm. Insets are zoom in images of corresponding surfaces. Scale bars: 10 μm

are generated, indicating some material vaporizes when the pulse overlap increases. When the overlap ratio further increases, gray surface is achieved and the grooves grow much deeper. More fluffy pores can be seen when the magnification is further enlarged. Pure black aluminum is obtained with an extremely high pulse overlap, which surface shows deep rill-like folds and the grooves can be as deep as 125 μm according to our measurement. Moreover, some other types of structures appear, such as nano-voids, protrusions and flower-like aggregates. Even though the spatial periods are pronounced as we see from each SEM image, all the colors remain the same when changing the viewing angle. The little angle-dependence suggests that the isotropic nano- or submicro-structures play a dominant role in the coloring effect rather than the microscale periodic structures. Iridescent colors can also be produced by the femtosecond laser modification by slightly adjusting the processing parameters to make sure the inscribed periodic structures to predominate in light diffraction, analogous to a two-dimensional grating. Our further studies reveal that not only the surface topography but also the chemical compositions are changed by the laser treating, which may also affect the color exhibition [1].

4.2 Polymer Replicas of Metal Masters and Their Wetting Properties

Laser modifying metal can be used to produce the templates, much more competitive than ultraviolet lithography or electron-beam lithography due to its low cost and simply equipped. It is reasonable to assume that the replicated surfaces possess the same function as the main structure features can be precisely transferred. The replicated PDMS film can be easily peeled off from the master and the desired surface structure is obtained, as shown in Fig. 4.3.

Figure 4.4 shows the SEM images of the PDMS surfaces replica of the different color aluminum [1]. It can be seen that the structures are successfully transferred, thus complementary structures are obtained. For example, the peak lines in gray aluminum are narrower than the valleys; correspondingly the peaks in PDMS replica become very wide and relatively flat with narrow gaps between them. In our experiment, when the pulse overlap is too high (>97%), the transfer process become incompetent as the depths of grooves increase and the liquid PDMS is difficult to penetrate into them. As shown in Fig. 4.4d, some smooth regions exist among the replicated structures caused by the residual polymers which should seep into the grooves. Figure 4.4e, f show the magnified details of the gray and black aluminum surface replicas, indicating that

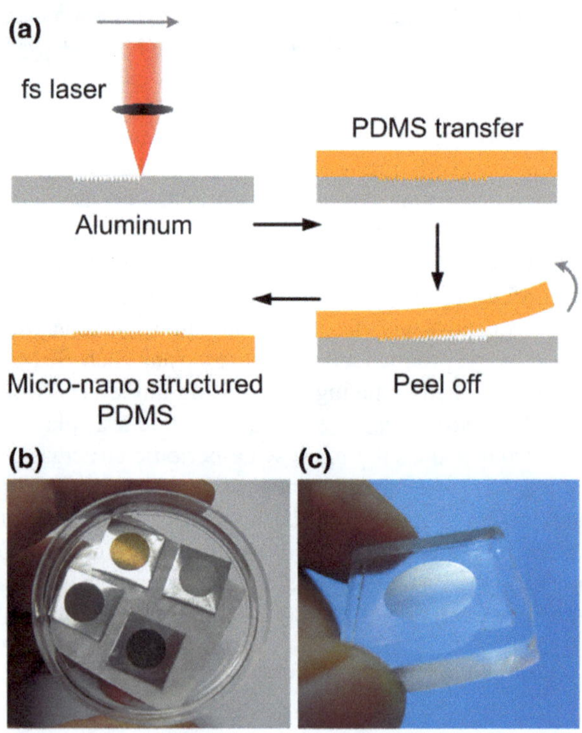

Fig. 4.3 **a** The schematic illustration of the fabrication process of micro/nanostructured polymer; **b** the solidified PDMS covered on the colorful aluminum templates; **c** the photograph of PDMS after being peeled off

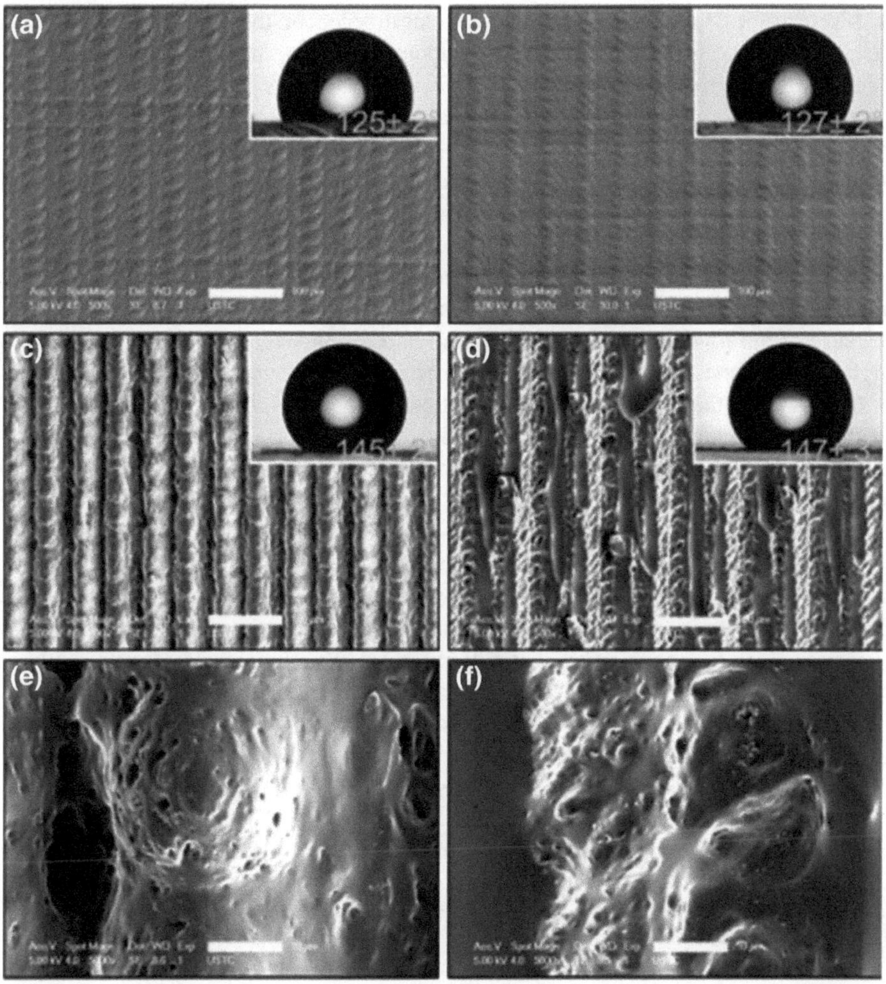

Fig. 4.4 **a**–**d** SEM images of replicated PDMS by color aluminums. Scale bars: 100 μm. The insets are the micro-droplets on the corresponding surfaces; **e** and **f** are magnified images of **c** and **d**. Scale bars: 10 μm

nano-voids and protrusions are successfully formed on PDMS surfaces. It is worth to be noted that the metal masters can be reused, giving the method decided advantages of good repeatability and cost-efficient. According to our experimental observation, no colors appear on the replicated surfaces although the structures are successfully transferred. It indicates that not only the hierarchical micro/nano-structures, but also other material surface properties, such as refractive index and reflectivity affect the color exhibition. However, besides structural colors, other surface functions, like water repellent, can be exploited in the duplicated PDMS surfaces.

Insets of Fig. 4.4a–d show the sessile water droplets on the structured PDMS surfaces [1]. For the surface replicated by golden aluminum, the CA is 125° ± 2°, distinctly bigger than a flat PDMS, which is measured to be 109° ± 1°. The CAs increase for the white, gray and black aluminum transferred surfaces, which are measured to be 127° ± 2°, 145° ± 2°, and 147° ± 3°, respectively. Hence the CAs increase monotonically with the pulse overlap. Here, the contact model can be explained by Wenzel and Cassie-Baxter states. Wenzel model describes homogeneous wetting and represents a wet-contact mode of water and rough surface, where water droplets pin the surface to form high CA hysteresis. Cassie model describes heterogeneous wetting and represents a non-wet-contact mode where water droplets can roll off effortlessly because of the low CA hysteresis. Generally speaking, Wenzel's state can induce a high CA hysteresis. We can see that the apparent CAs change with the surface roughness not matter the droplets are in Wenzel or Cassie cases. In fact, according to the studies on the natural hydrophobic adhesive leaves and the biomimicking preparations, the structured surfaces here are in Wenzel or meta-stable Wenzel/Cassie hybrid states, which make them exhibiting a CA larger than the critical CA (an angle above which the wetting type is thought to be a stable Cassie type) and a high CA hysteresis. As the pulse overlap gets higher, deeper grooves are directly fabricated and more secondary structures, such as nano-particles, voids and other aggregates, are induced, resulting in rougher surfaces of the aluminum templates. For rough surfaces especially with two-scale micro/nano hierarchical structures, water droplet will not invade into the microscopic structures totally and some air exists between solid and liquid phase, contributing to the phenomenon that the apparent CA of the heterogeneous surface is obviously greater than the intrinsic CA of a flat smooth surface. As mentioned above, when the pulse overlap is too high, the replicated structures get some undesired defects, where should be sharp crests. That makes more possible solid fraction contact the water droplet, therefore the differences between the CAs shown in the insets of Fig. 4.4c, d are not so distinguishable.

4.3 Tunable Contact Angle and Strong Adhesion Force

As the CA is associated with the laser pulse overlap, a facile approach to produce surfaces with controllable wetting properties can be derived. The involved laser texturing technique is much simpler and more versatile than other micro/nano fabrication methods and the pulse overlap adjustment is easy and proficient. Furthermore, soft lithography facilities the mass production of large-area structured surfaces. In order to demonstrate the feasibility, a series of PDMS surfaces are produced by the metal masters treated with different femtosecond laser pulse overlaps, as shown in Fig. 4.5. The inset shows the corresponding shape of water droplets on the replicated surfaces. It can be seen that the CAs increase monotonically with the pulse overlap, ranging from 126° to 147°. For the case of pulse overlap is less than 60%, the increase of CAs is almost linear and the values vary gently (the region left to the dashed dotted line). When the pulse overlap is further increased, the changes of CAs

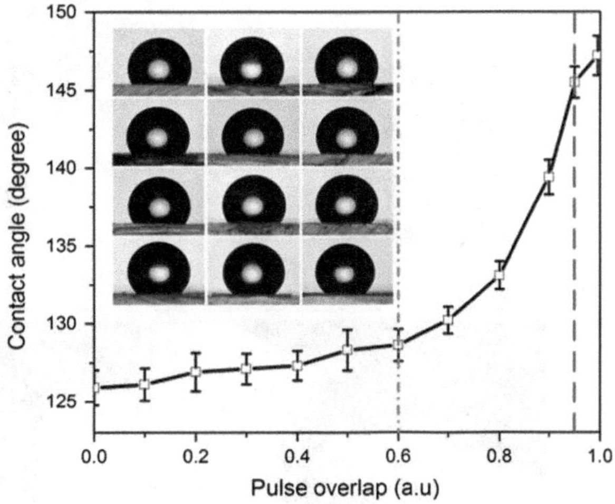

Fig. 4.5 The dependence of contact angle on the pulse overlap, changed from 0 to 99.5%. The inset is the shape of the water droplets on different surfaces

become rapid following an approximate exponential law (the region between the dashed dotted line and the dashed line). When the pulse overlap is larger than 95%, the increase of CAs becomes slow as mentioned above. As we can see, the CAs are all less than 150°, beyond which the surface is so-called "superhydrophobic". The surface roughness and CA can be further enhanced by other means, like chemical coating (e.g., fluoroalkylsilane) to lower the surface free energy or electroless plating to deposit nanoparticles on the primary structures, which is widely conducted and requires minimal space here.

Natural rose petal exhibits special hydrophobic and adhesion property, different from the lotus self-cleaning surface on which a droplet spontaneously rolls off with a slight tremble. Jiang et al. firstly studied the mechanism of rose petal and imitated the "sticky" surfaces by duplicating the real petal's surface and fabricating aligned polystyrene nanotubes. Since then, many researchers devoted themselves to the field of artificial rose-petal-like surfaces preparation. Compared to the lotus leaf, the size of the hierarchical micro/nano-structures on the rose petal is relatively larger, and water droplet can enter into larger grooves, thus giving rise to a higher water penetration depth and a resultant higher adhesive force. High adhesive force between solid surface and liquid causes visible great CA hysteresis. The mesoscale period of the PDMS surface structures is 50 μm and all the laser textured templating surfaces show sticky properties due to the great CA hysteresis. Take the black aluminum duplicated polymer film as an example, the water drop stay pinned steadily to the vertical surface, as shown in Fig. 4.6a. The water drop is also adsorbed and not rolling off from the surface when we tilt the sample at different angles (135° and 150° as shown in Fig. 4.6b, c) or even turn it upside down (Fig. 4.6d). This sticky hydrophobic characteristic facilitates many potential applications like tiny water droplets transportation without loss and analysis of small volumes of liquid.

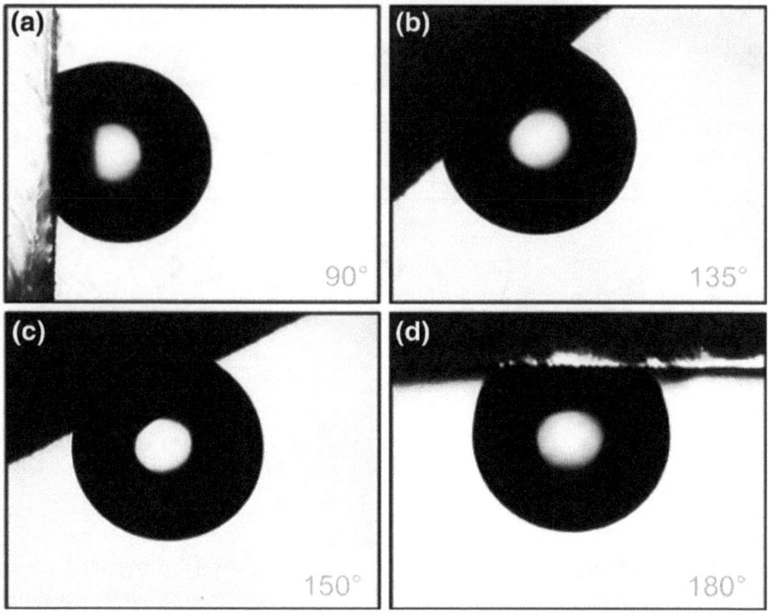

Fig. 4.6 Behavior of water droplets on the as-prepared PDMS surfaces at different tilt angles of **a** 90°, **b** 135°, **c** 150°, and **d** 180°

In summary, femtosecond laser textured templating method is applied to fabricate functional hydrophobic surfaces with a high adhesive force. Amplified femtosecond laser is employed to functionalize aluminum with diverse colors including golden, white, gray and pure black. It reveals that isotropic nano/submicro-structures dominate the color effect rather than microscale periodic structures. Subsequently the color metals are used as templates for soft lithography process, in which structured polymers are readily prepared with controllable wetting properties. The contact angle can be finely tuned from 125° to 147°. Furthermore, the as-prepared surfaces exhibit a high adhesive force to water drops due to the deep penetration depth and the resultant great CA hysteresis. This approach features simplicity, rapidness, and easiness of large-area fabrication, which can find extensive applications in various flexible functional surfaces manufacturing [1].

Reference

1. Hu Y, Li G, Cai J, et al. Facile fabrication of functional PDMS surfaces with tunable wettablity and high adhesive force via femtosecond laser textured templating. AIP Adv. 2014;4(12):127141.

Chapter 5
Multifunctional Three-Dimensional Porous Metal Micro/Nanocages by Ethanol-Assisted Femtosecond Laser Irradiation

Metallic micro/nanostructures with controllable 2D-3D geometries have been attracting great attention due to their many promising applications ranging from plasmonics [1, 2], electronics, [3] and bioscience [4] to chemistry [5, 6]. Significant effort has been devoted to the exploration of effective approaches for realizing various controllable metallic micro/nanostructures, investigating their performance, and exploring their novel applications. For example, conventional top-down microfabrication methods containing electron beam lithography [7], focused ion beam (FIB) [8], and X-ray lithography [9] have been widely used in metal micro/nano structures fabrication due to their high resolution, but how to effectively and economically produce such nanostructures with large area is still a big challenge. On the other hand, bottom-up methods [10–13] by combining colloidal microsphere self-assembly, metal deposition and lift-off processing were usually used for fabricating 2D period metal microstructures. In addition, an advanced two-photon polymerization (TPP) technique [14, 15] combined with a site-selective metal deposition technique has been developed for fabricating more complex structures with ~100 nm resolution, which are useful in some specific applications. However, the process is complicated, and the area is usually up to several hundred square micrometers. Moreover, many of them are limited for general use such as microelectronics, plamonics, and nanophotonics due to the nonconducting polymeric structures. Thus, it is highly desirable to develop a convenient technology that is able to rapidly fabricate large-area 2D-3D micro/nanostructured metallic surfaces with excellent performance.

Recent studies have shown that self-organized metallic micro/nanostructures can be created on metal surfaces using femtosecond laser direct irradiation. Among these structures, the most common ones are the subwavelength ripples, [16–18] nanotextured conical microstructures, [19] jet like structure, [20] which can greatly improve the optical, biological, and chemical properties of metallic material surface, such as color display, anti-counterfeiting, and optical data storage. Nevertheless, these structures are mostly limited to one-dimensional (1D) or two-dimensional (2D) features, which seriously limit their multifunctionality [16–20]. As a kind of important ferromagnetic metal, nickel shows outstanding optical, catalytic, electrical, and superior resistant characteristics to corrosion, which makes it widely applicable in electronics, nickel-batteries, alloy, and chemical catalyst [21–38]. Although some functional

G. Li, *Bionic Functional Structures by Femtosecond Laser Micro/nanofabrication Technologies*, Springer Theses, https://doi.org/10.1007/978-981-13-0359-3_5

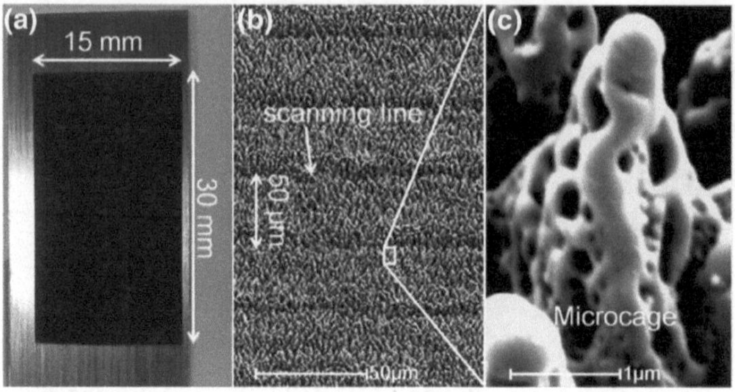

Fig. 5.1 Large-area fabrication of porous 3D micro/nanocage-like arrays on nickel surface by one-step ethanol-assisted femtosecond laser irradiation

1D-2D structures such as ripples, mounds, and cones have been formed on nickel surfaces, there are still few reports on the controllable fabrication of complex 2D-3D micro/nanostructures and systemic investigation of their multifunctions.

Here, we report a facile and rapid method to fabricate large-area (4 cm^2 or more) (Fig. 5.1) three-dimensional (3D) porous metal micro/nanocages arrays with 0.8 μm sizes by one-step ethanol-assisted femtosecond laser irradiation on nickel surfaces [38]. The underlying formation mechanism was investigated by a series of femtosecond laser irradiation under exposure time from 5 to 30 ms. By controlling the laser energy, micro/nanocage arrays with different sizes from 0.8 to 2 μm were obtained. Furthermore, these porous micro/nanocages arrays exhibit multiple functions, which can endow the nickel surfaces with not only low reflectivity over a broad wavelength range but also improved hydrophobic ability.

5.1 Experimental Section

To fabricate micro/nanocage-like structures, the mechanically polished nickel plates (Chinese New Metal Materials Technology Co., Ltd., China) with purity of 99.99% and dimension of 0.5 × 20 × 20 mm^3 are chosen as targets. The nickel plate is stuck to the bottom of a 20 mL container (Petri dish), which is filled with ethanol (purity of 99.3%). The diameter of the focused spot is approximately 30 μm. The exposure time of the laser beam on the sample surfaces is controlled by a computer-controlled high-speed shutter. To prepare porous micro/nanocages, the experimental conditions were precisely controlled and optimized. The incident laser pulse energies are set from 0.05 to 0.20 mJ/pulse. The step size between the adjacent scanning lines is 50 μm, and the total scanning time is ~10 min. During the fabrication process, the CCD camera was connected to computer for clear online observation in nickel pattern surface. In contrast, the nickel plates are also stuck to the bottom of the empty Petri dish, or it is filled with distilled water for laser irradiation with the same parameters as in ethanol. To quantitatively characterize the spectral responsivity, the wavelength dependence of total reflectance in the UV, visible, and IR regions (250–2500 nm) was measured by a spectrophotometer (SOLID3700, Shimadzu, Japan) with an integrating sphere, which can effectively collect the stray lights by the diffuse reflections. The wettability was investigated by dripping a 4 μL water droplet on the processed nickel surfaces to perform the contact angle (CA) measurements with contact-angle goniometer (CA100C, Innuo, Shanghai, China) by a sessile drop method. The average CA values were obtained by measuring the same samples in three different positions. The digital drop image taken by the digital camera was processed by the software (CA100C, Innuo, Shanghai, China) automatically, which can calculate the contact angles from the drop shape.

5.2 Formation of 3D Micro/Nanocage-like Structures

From Fig. 5.2a, b, it is clearly indicated that different microstructures are formed on the femtosecond laser radiated surface of nickel under different ambient environments [38]. The laser energy is the same, ~0.05 mJ/pulse. We found that the formed structures in air and in water are 2D with 0.02–0.3 μm and 0.03–0.4 μm size, while the ones in ethanol look like 3D micro/nanocage with 0.8 μm size. The comparison of air, distilled water, and ethanol ambient environments clearly indicates the unique 3D micro/nanocage-like structures can only be formed in ethanol (Fig. 5.2c). The formation of the micro/nanocage arrays by ethanol-assisted femtosecond laser irradiation raises a fundamental question: how are the structures induced? Compared to the irradiation in gaseous environment, the formation process in ethanol is more complex due to the effect by laser-induced plasma of the ambient liquid, enhanced heat conduction, intensified acoustic pressure, increased shock wave, and explosive vaporization. The formation of unique structures may be caused by the competition of several processes since the phase transition occurs both in the solid and adjacent liquid. Femtosecond laser may produce the superheated substance and a transient zone of elevated pressure near the target, which brings the surrounding liquid to a supercritical state. As a result, the pressure wave interacts with the melt layer on the target surface, and its morphology is perturbed. This surface profile is frozen upon cooling by the subsequent liquid to prevent being smeared by the surface tension of the melt [24–26].

The scanning of laser beam across the nickel surface ensures the uniform distribution of laser irradiation along the scanning line. The morphology of micro/nanocages depends strongly on the pulse energy and the irradiation time. In our experiment, the micro/nanocages arrays can be fabricated at scanning speed of 1 mm/s, even with the low pulse energy of 0.05 mJ/pulse. Therefore, the average irradiation time on the excitation spot is 30 ms, as shown in Fig. 5.3a. To study the formation and evolution of the micro/nanocages, the scanning speed was set as 6 mm/s, and the average irradiation time on the excitation spot is 5 ms, as shown in Fig. 5.3b. Then, the processing

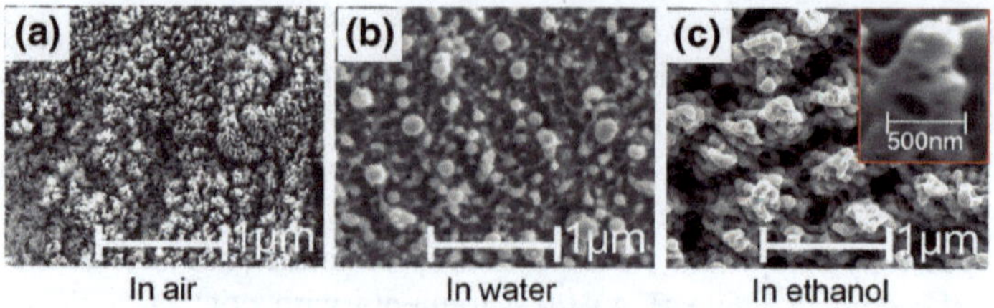

Fig. 5.2 **a** SEM image of the nickel surface irradiated in air. The laser-induced microstructures are irregular; **b** Spherical microstructures formed by laser irradiation in distilled water; **c** Unique 3D micro/nanocage-like structures fabricated in ethanol solvent. The inset is 45° tilted SEM image of single micro/nanocage-like microstructures. The size of micro/nanocage is ~0.8 μm, and the average size of its surface nanohole is ~10–100 nm

Fig. 5.3 In situ evolution process of 3D micro/nanocage-like structures

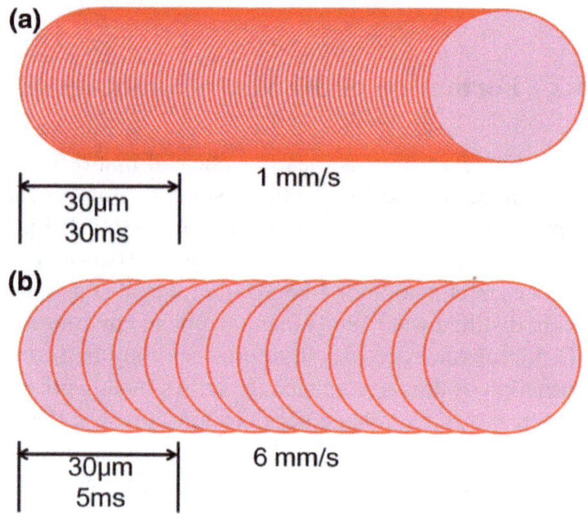

area was scanned for different times from 1 to 6, and the corresponding irradiation time on the excitation spot was 5, 10, 15, 20, 25, and 30 ms, respectively. After each radiation, the morphology was observed by SEM images, as shown in Fig. 5.4. Through this method, the formation mechanism was systemically investigated [38].

The evolution with increasing irradiation time reveals a clear formation process for the micro/nanocage-like structures. First, the laser induces plasma plume on the ethanol/nickel interface and ablates the nickel surface to form initial roughness, as shown in Fig. 5.4a. Then, with the increasing irradiation time from 10 to 30 ms, the confined laser-induced plasma by ethanol can further interact with the formed structures, which results in the growing structures (Fig. 5.4b–f). Because of the confinement of the pure ethanol, the laser-induced plasma can be pushed to a state of high temperature, high pressure, and high density. Finally, the porous metal structures

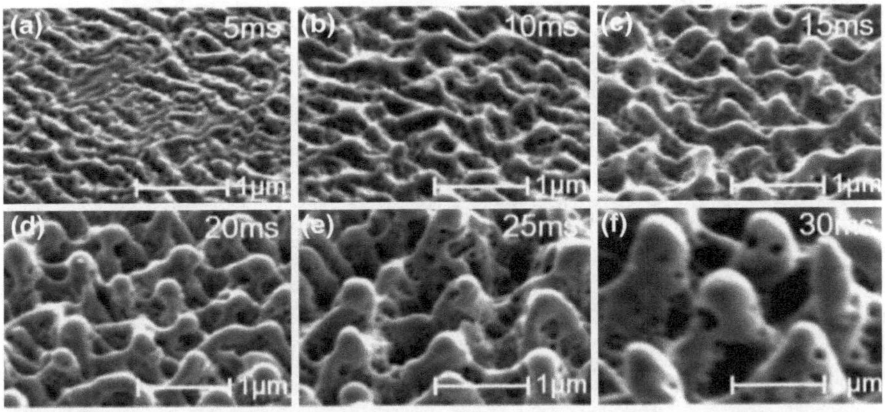

Fig. 5.4 45° tilted SEM images of nickel surface structures irradiated with different exposure time from 5 to 30 ms

are formed. Additionally, the cooling effect of the confined ethanol on the laser-induced plasma also enhances the formation of the porous microstructures [38].

5.3 Precisely Controlling the Micro/Nanocages Size by Adjusting the Laser Pulse Energy

On the basis of the evolution results, we also demonstrate the growth control of the micro/nanocage-like structures by adjusting the pulse energy. Figure 5.5a–d indicates various porous micro/nanocage-like structures prepared under pulse energy ranging from 0.05 to 0.20 mJ/pulse at scanning speed of 1 mm/s. From the magnified SEM images in Fig. 5.5e–h, it can be seen that the height, size, and spacing of micro/nanocages depends on the pulse energy. As the pulse energy increases, the micro/nanocages become larger due to the stronger interaction of laser with nickel surfaces. At low pulse energy of 0.05 mJ/pulse, the laser can only induce micro/nanocage with size as small as 0.8 μm. With the increase of pulse energy, the laser-induced plasma is further expanded to produce a shock wave, higher temperature, and higher pressure at the nickel surface, so the higher and larger size of micro/nanocages are produced.

Judging from the SEM images, it is found that the distribution of these micro/nanocages is uniform and exhibit periodic. To further investigate the distribution of the micro/nanocages, the top-view SEM images of Fig. 5.5a–d are subjected to a two-dimensional fast Fourier transformation (2D-FFT), which can be expressed as Eq. 5.1 [27, 28].

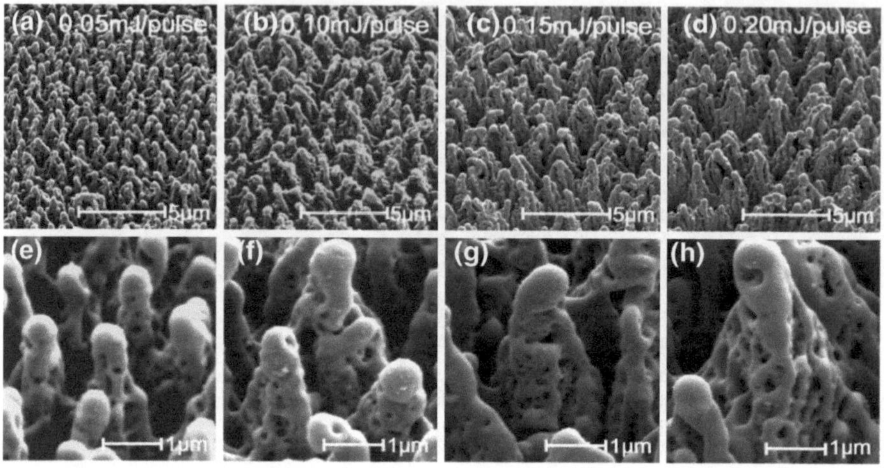

Fig. 5.5 Precisely controlling the micro/nanocages size by adjusting the laser pulse energy. **a–d** SEM images of various 3D micro/nanocages arrays produced at pulse energy of 0.05, 0.10, 0.15, and 0.20 mJ/pulse, respectively. The scanning speed is 1 mm/s. **e–h** Magnified SEM images of micro/nanocages arrays with various sizes (0.82, 1.12, 1.54, and 1.91 μm)

$$
\begin{aligned}
F\{g(x, y)\} &= \int_{-\infty}^{\infty}\int_{-\infty}^{\infty} g(x, y)\exp\left[-j2\pi(f_X x + f_Y y)\right]dxdy \\
&= \int_{-\infty}^{\infty} g_X(x)\exp\left[-j2\pi f_X x\right]dx \int_{-\infty}^{\infty} g_Y(y)\exp\left[-j2\pi f_Y y\right]dy \\
&= F_X\{gx\}\, F_Y\{gy\}
\end{aligned}
\tag{5.1}
$$

where *x*, *y* represent the rectangular coordinates, g and f are the time domain and frequency domain function, respectively. *X*, *Y* are the subscript, and *gX(x)*, *gY(y)* are the function factorized by *g(x, y)*, respectively. In our case, *gX(x)*, *gY(y)* represent the distribution in *x*, *y* directions.

The transformation results is shown in Fig. 5.6e–h, which reveals that frequency spectrum is distributed into ring, and the radius of the ring is diffuse within a small range. This demonstrates that the micro/nanocages are formed with characteristic periodic distribution in all directions. Furthermore, the radius of the ring can reflect the denseness of the distribution. This can be confirmed from Fig. 5.6e–h, from which we can see that the diameters of the rings are becoming smaller and smaller. The larger the diameter, the more intensive the micro/nanocages.

Figure 5.7 shows typical size distribution histograms of the micro/nanocages, which are based on the statistical calculations of all SEM analysis. From Fig. 5.7, we can see that the sizes from 0.79 to 0.84 μm and from 1.09 to 1.13 μm almost occupy 80% of the micro/nanocages, respectively. Although the distributions are slightly in disorder, the frequency of the sizes from 1.50 to 1.58 μm and from 1.86 to 1.97 μm is also more than 80% (Fig. 5.7c, d). Generally, the sizes of the micro/nanocages are still uniform, and the average size increases with the increased pulse energy.

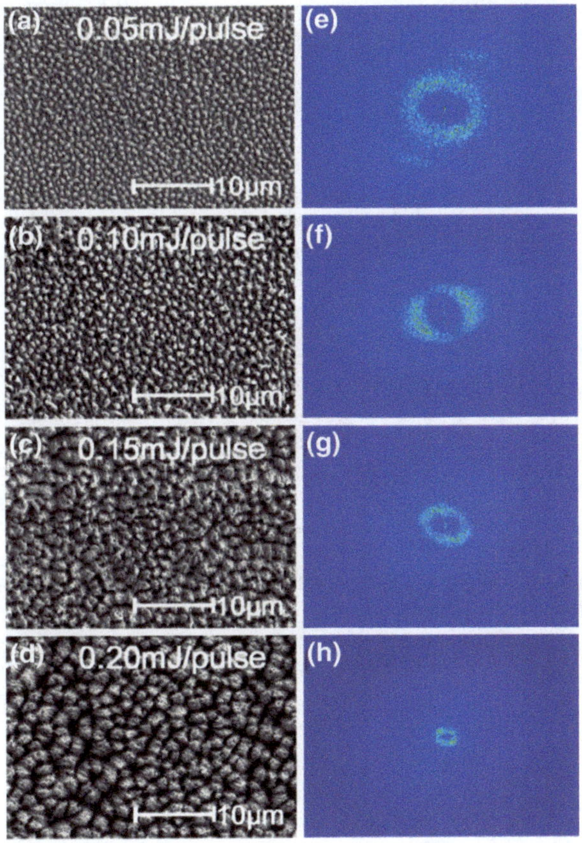

Fig. 5.6 Quantitative analysis of the size distribution of various micro/nanocages. **a–d** Top view SEM images of the micro/nanocages produced at 0.05, 0.10, 0.15, and 0.20 mJ/pulse, respectively. **e–h** The 2D fast Fourier transformation (2D-FFT) results for the SEM images

Finally, from the statistical results in Fig. 5.8a, b, it is revealed that the heights of the micro/nanocages increased, while the density (the number of the micro/nanocages per square centimeter) decreased with the increasing pulse energy [38]. What's more, plenty of tiny nanoholes with the size from tens to hundreds of nanometers are formed over the entire micro/nanocage. To the best of our knowl-

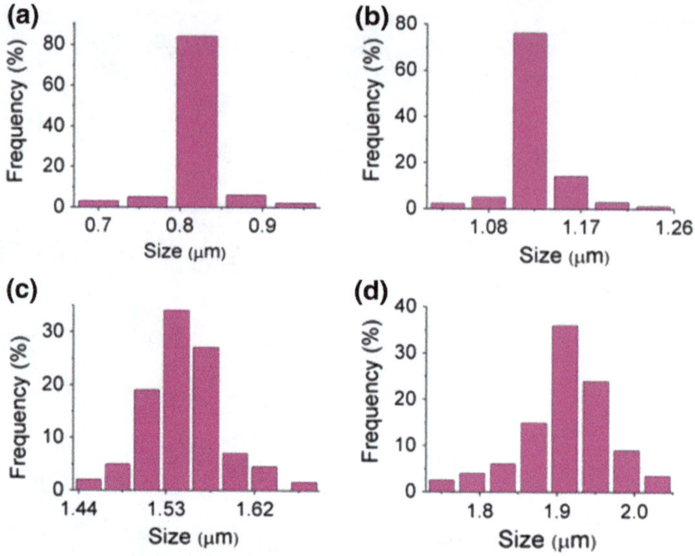

Fig. 5.7 The size distribution histograms of the micro/nanocages based on the statistical calculations of all the SEM images

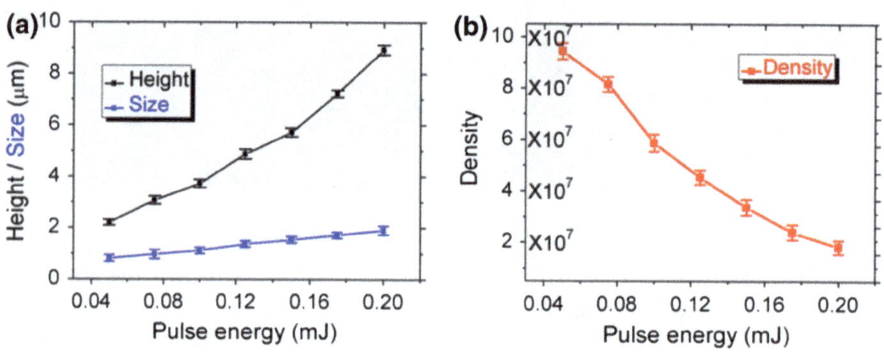

Fig. 5.8 **a** The height and the size of the micro/nanocages as a function of the pulse energy; **b** The density of the micro/nanocages with different pulse energy

edge, the 3D multiscale porous micro/nanostructures prepared on nickel have not been reported. Our proposed method also shows the ability to control the size of the micro/nanocages by adjusting the laser energy.

In the experiment, the pulse energy employed for the control of the growth of the micro/nanocage is chosen in the range from 0.05 to 0.20 mJ/pulse and the scanning speed is set as 1 mm/s. It is found that the formed micro/nanocage is not perfect if the energy is too big (>0.2 mJ) or too small (<0.05 mJ). Figure 5.9a SEM image of micro/nanostructures produced on nickel surface with pulse energy of 0.04 mJ/pulse. One can see that typical columnar microstructures with spherical structures at the top

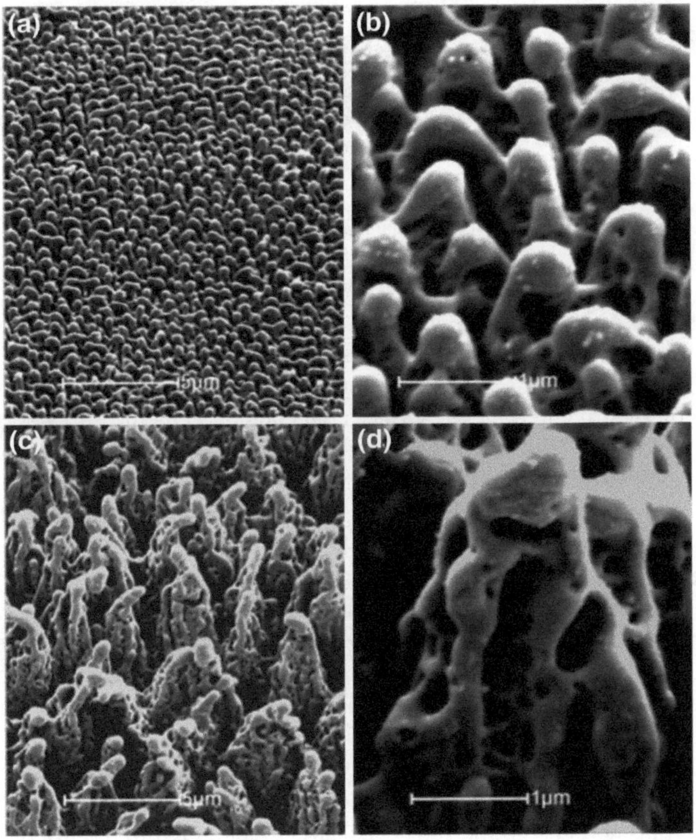

Fig. 5.9 Irradiating nickel surface with pulse energy lower than 0.05 mJ or greater than 0.20 mJ

are formed. The magnified SEM image in Fig. 5.9b shows that columnar microstructures exhibit porous, and look like "squid". In this case, the holes are small and the less. Figure 5.9c SEM image of the imperfect micro/nanocages which are produced at the pulse energy of 0.25 mJ/pulse. It is seen that the spindly micro/nanocages are no longer upright. Most of the heads are hung. A careful observation from the magnified SEM image Fig. 5.9d demonstrates that most of the round holes are broken into long cracks. This kind of micro/nanocages look like 3D skeletons [38].

5.4 Enhanced Antireflection of 3D Porous Metal Micro/Nanocages

From the digital photographs (Fig. 5.10) [38], it is seen that the processed nickel surfaces show black color [38]. The nickel surfaces reflectance before and after

Fig. 5.10 Photograph of the polished and black nickel

femtosecond laser irradiation is also measured (Fig. 5.11) [38]. It is evident that the reflectance of the black nickel surface is much lower than that of the polished one (~80%) over the entire wavelength range (0.25–2.5 μm). Furthermore, the reflection of structured nickel surfaces depends upon the height and spacing of the micro/nanocages. The larger the micro/nanocages, the lower the light reflection. Note that the black nickel produced with the pulse energy of 0.20 mJ/pulse only has less than 7% reflection in the entire wavelength range. This novel phenomenon is presumed to be related to the tapered morphology of the micro/nanocage and surface plasmon excitation. First, the tapered micro/nanocages can form effective graded refractive index layer at the air/solid interface and significantly reduce the reflection. Second, the porous structures may form photonic traps and enable the incident light to be confined [29–31]. Third, it was found that the lateral of the micro/nanocages with larger sizes were covered with many smaller-size nanoparticles and nanoholes. So, the incident light was effectively absorbed and multiply reflected by these nanoparticles and nanoholes due to surface-plasmon absorption. Finally, successive multiple reflection may take place between/inside the micro/nanocages, which further enhance the surface-plasmon absorption effect. All these factors led to the lower reflection over a wide wavelength of 250–2500 nm. The angle-dependent reflectance, for example, the micro/nanocages with size of 1.91 μm, was also investigated with incident angle of 6° and 12°. The measurements demonstrate that the angle has little impact on the reflectance (Fig. 5.12) [38]. Although the reflectance from 1250 nm is higher than that of the normal-incidence reflection, it is still less than 10% in the range from 250 to 1250 nm. Moreover, the micro/nanocages are conductive due to the nickel substrate, so the electrical properties of these structures are very useful in optoelectronics, electronics, industry, and batteries [21, 23, 32].

5.5 The Wetting Property of 3D Porous Metal Micro/Nanocages

It has been reported that the laser-machined micro/nanostructures have the ability to tune surface wettability, making the metal surface more hydrophilic or hydrophobic. To investigate the wettability, the CA measurements on the processed nickel surfaces were performed (Fig. 5.13) [38]. The wettability of the polished nickel is also measured for comparison. The nickel with clean surfaces is initially hydrophilic;

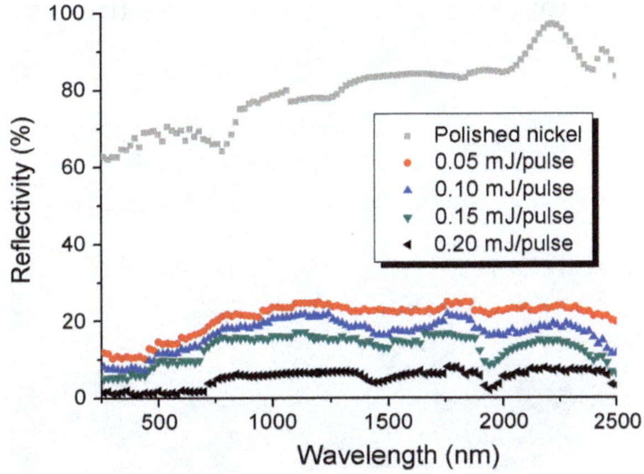

Fig. 5.11 Wavelength-dependent reflectance of polished and black nickel

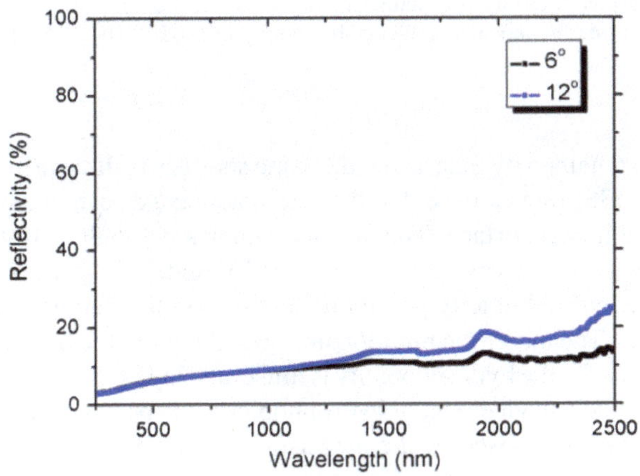

Fig. 5.12 The angle-dependent reflectance measurements

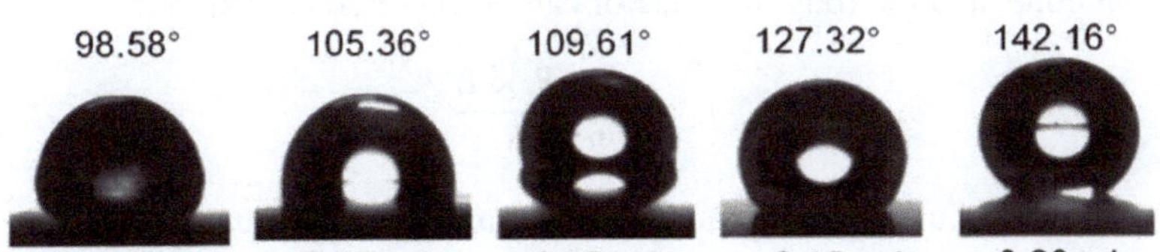

Fig. 5.13 The CA of the polished and black nickel without any low surface energy modification

however, the polished nickel in our case shows hydrophobic due to the nickel oxidation in air. It is well-known that the nickel is not stable in air and easily reacts with oxygen to form nickel oxide [23, 33]. So, the CA is increased to 98°.

The CA value was measured after the droplets were dripped onto the samples for 5 s, and there was no scattering of contact angles along the surfaces. The average

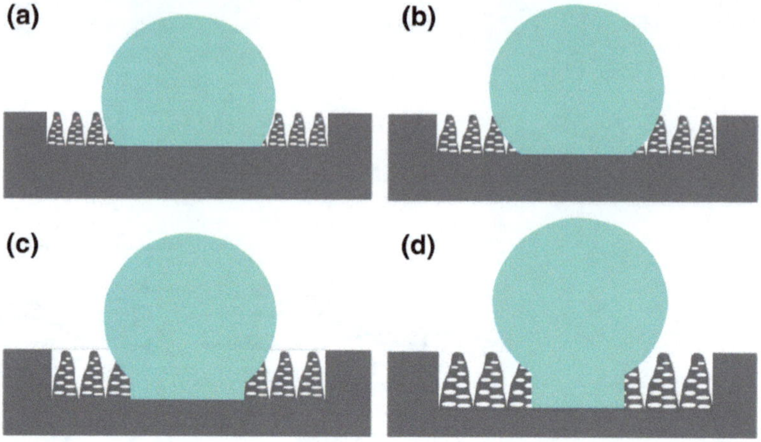

Fig. 5.14 The analysis of the wetting state between the water droplet and the samples, which are covered by various micro/nanocages under pulse energy of 0.05, 0.10, 0.15, and 0.20 mJ/pulse, respectively

CA values were obtained by measuring the same samples at three different positions. From Fig. 5.13, [38] one can see that the micro/nanocages can enhance hydrophobic ability of the nickel surface. The contact angles are 105.36°, 109.61°, 127.32°, and 142.16° for pulse energy of 0.05, 0.10, 0.15, and 0.20 mJ, respectively. It is obvious that the hydrophobicity greatly depends upon the heights and sizes of the micro/nanocages. The larger the micro/nanocages, the more the hydrophobicity. The schematic diagram for the hydrophobicity is illustrated in Fig. 5.14 [38]. According to Wenzel's model, the enhancement of hydrophobicity can be attributed to the increase of the roughness factor caused by the increasing heights of the micro/nanocages.

Here the roughness factor for a micro/nanocage is defined as the ratio of the lateral area to the projected area. The micro/nanocage model can be described by an equivalent cone, and the roughness factor can be expressed as Eq. 5.2:

$$f = \frac{\pi \times R \times L}{\pi \times R^2} \tag{5.2}$$

where R is the radius of micro/nanocage bottom, which is equal to the size of the micro/nanocage. L is the generatrix of the micro/nanocage. On account of

$$L = \sqrt{H^2 + R^2} \tag{5.3}$$

where H is the height of the micro/nanocage. The roughness factor formula can be rewritten as Eq. 5.4

$$f = \sqrt{1 + \left(\frac{H}{R}\right)^2} \tag{5.4}$$

Table 5.1 The measured R, H, $\varepsilon(R)$ and $\varepsilon(H)$ for the micro/nanocage arrays

Pulse energy	0.05	0.075	0.10	0.125	0.15	0.175	0.20
$R(\mu m)$	0.82	0.97	1.12	1.37	1.54	1.72	1.91
$\varepsilon(R)(\mu m)$	±0.13	±0.14	±0.12	±0.12	±0.13	±0.12	±0.17
$H(\mu m)$	2.21	3.07	3.73	4.88	5.72	7.23	8.94
$\varepsilon(H)(\mu m)$	±0.11	±0.13	±0.15	±0.19	±0.17	±0.14	±0.19

Table 5.2 The calculated f and $\varepsilon(f)$ for the micro/nanocage arrays

Pulse energy	0.05	0.075	0.10	0.125	0.15	0.175	0.20
f	2.87	3.32	3.48	3.70	3.85	4.32	4.77
ε	±0.53	±0.56	±0.47	±0.43	±0.41	±0.36	±0.51

The error can be expressed as Eq. 5.5:

$$\begin{aligned}\varepsilon(f) &\approx \left(\frac{\partial f}{\partial R}\right) \bullet \varepsilon(R) + \left(\frac{\partial f}{\partial H}\right) \bullet \varepsilon(H) \\ &= \left(1 + \left(\frac{H}{R}\right)^2\right)^{-\frac{1}{2}} \bullet \left(\frac{H}{R}\right)^2 \bullet \left[\frac{\varepsilon(H)}{H} - \frac{\varepsilon(R)}{R}\right] \\ &\leq \left(1 + \left(\frac{H}{R}\right)^2\right)^{-\frac{1}{2}} \bullet \left(\frac{H}{R}\right)^2 \bullet \left[\left|\frac{\varepsilon(H)}{H}\right| + \left|\frac{\varepsilon(R)}{R}\right|\right]\end{aligned} \tag{5.5}$$

The measured R, H, $\varepsilon(R)$ and $\varepsilon(H)$ are shown in Table 5.1.

We can obtain f and $\varepsilon(f)$, as shown in Table 5.2.

The roughness factor based on the data in Fig. 5.8a is calculated, as shown in Fig. 5.15 and, Tables 5.1 and 5.2 [38]. It is indicated that the roughness factor increased with the increasing micro/nanocage height. The biggest f can reach 4.77, which is much bigger than the one of flat surface ($f = 1$). In addition, the nanoholes (~100 nm) on the micro/nanocages also increase the superficial area of micro/nanocages, which further enhance the roughness factor. The maximum contact angle can reach as much as 142° without additional chemical modification. The contact angle may be over 160° if the surface is modified with low surface-energy materials [34–37]. The dust particles and any pollutants on the hydrophobic surface can be clearly washed by the water droplets (Fig. 5.16) [38].

To investigate the durability of the micro/nanocages (Fig. 5.17) [38], we load the structures surface by a weight of 200 g and then scrub it slightly five times. The SEM results show that the micro/nanocages structures are well-preserved. In addition, the water CA is 141.01°, almost no change compared with the previous one.

Fig. 5.15 The roughness factors calculated from the data in Fig. 5.8 **a**. The biggest roughness factor can reach 4.77 for the microstructured nickel surfaces, which is much bigger than that of the flat surface ($f = 1$)

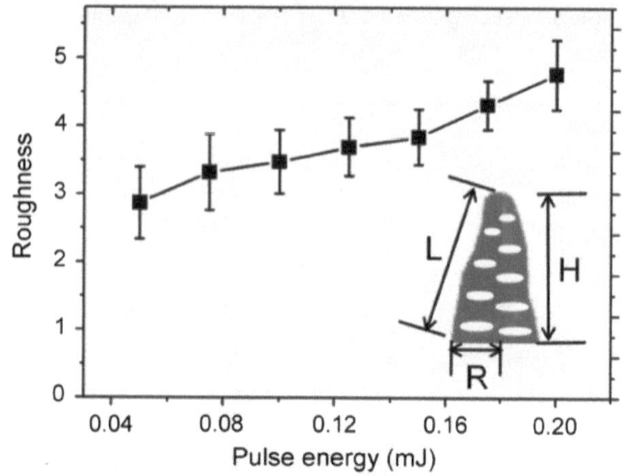

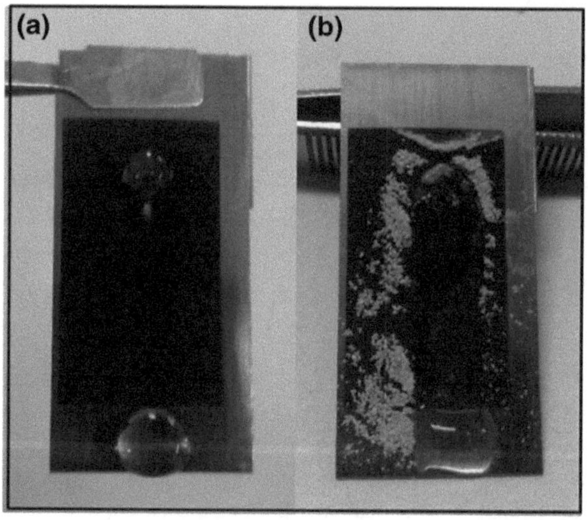

Fig. 5.16 **a** The water droplet rolls down along the almost superhydrophobic surface; **b** The dust particles and any pollutants on the superhydrophobic surface can be clearly washed out by the water droplets

In summary, large-area 3D porous metal micro/nanocages arrays are rapidly prepared on nickel surfaces by proposing a new method of one-step ethanol-assisted femtosecond laser irradiation. The formation process was in situ investigated by a series of femtosecond laser irradiation under exposure time from 5 to 30 ms, and its underlying mechanism is believed to be a combined process of accumulating pulse and confined plasmons. In addition, by tuning different laser energy from 0.05 to 0.2 mJ/pulse, the size of micro/nanocage arrays can be controlled from 0.8 to 1.9 μm. The 3D controllable cage-liked micro/nanostructures exhibit not only enhanced antireflection from 80 to 7% but also hydrophobicity with CA from 98° to

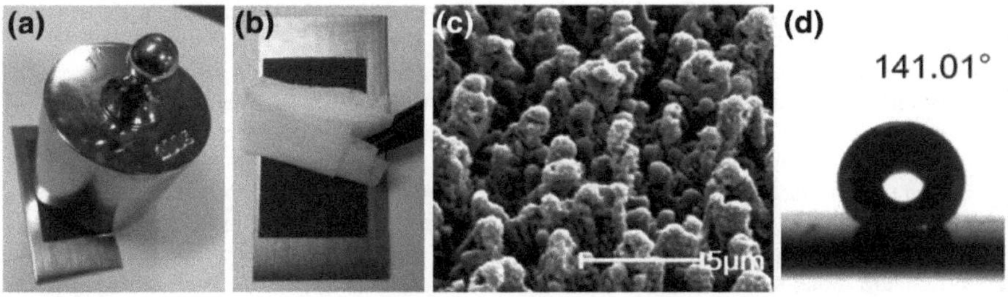

Fig. 5.17 The durability of the micro/nanocages with size of 1.91 μm in exploitation conditions

142° without surface modification. We believe that this simple and rapid method for 3D large-area controllable metal micro/nanostructures may find broader applications in optoelectronics, industry, batteries, and chemical catalysis.

References

1. Pile D. Photocatalysis: plasmonic enhancement. Nat Photonics. 2013;7(1):3.
2. O'Brien K, Lanzillotti-Kimura ND, Rho J, et al. Ultrafast acousto-plasmonic control and sensing in complex nanostructures. Nat Commun. 2014;5:4042.
3. Piglosiewicz B, Schmidt S, Park DJ, et al. Carrier-envelope phase effects on the strong-field photoemission of electrons from metallic nanostructures. Nat Photonics. 2014;8(1):37–42.
4. Shoji T, Saitoh J, Kitamura N, et al. Permanent fixing or reversible trapping and release of DNA micropatterns on a gold nanostructure using continuous-wave or femtosecond-pulsed near-infrared laser light. J Am Chem Soc. 2013;135(17):6643–8.
5. Linic S, Christopher P, Ingram DB. Plasmonic-metal nanostructures for efficient conversion of solar to chemical energy. Nat Mater. 2011;10(12).
6. Valsecchi C, Brolo AG. Periodic metallic nanostructures as plasmonic chemical sensors. Langmuir. 2013;29(19):5638–49.
7. Near R, Tabor C, Duan J, et al. Pronounced effects of anisotropy on plasmonic properties of nanorings fabricated by electron beam lithography. Nano Lett. 2012;12(4):2158–64.
8. Lin YY, Liao JD, Ju YH, et al. Focused ion beam-fabricated Au micro/nanostructures used as a surface enhanced Raman scattering-active substrate for trace detection of molecules and influenza virus. Nanotechnol. 2011;22(18):185308.
9. Pilet N, Raabe J, Stevenson SE, et al. Nanostructure characterization by a combined x-ray absorption/scanning force microscopy system. Nanotechnol. 2012;23(47):475708.
10. Ye X, Qi L. Two-dimensionally patterned nanostructures based on monolayer colloidal crystals: controllable fabrication, assembly, and applications. Nano Today. 2011;6(6):608–31.
11. Chang CH, Tian L, Hesse WR, et al. From two-dimensional colloidal self-assembly to three-dimensional nanolithography. Nano Lett. 2011;11(6):2533–7.
12. Tsujioka T. Metal-vapor deposition modulation on polymer surfaces prepared by the coffee-ring effect. Soft Matter. 2013;9(24):5681–5.
13. Gecys P, Markauskas E, Gedvilas M, et al. Ultrashort pulsed laser induced material lift-off processing of CZTSe thin-film solar cells. Sol Energy. 2014;102:82–90.
14. Masui K, Shoji S, Asaba K, et al. Laser fabrication of Au nanorod aggregates microstructures assisted by two-photon polymerization. Opt Express. 2011;19(23):22786–96.
15. Vora K, Kang SY, Shukla S, et al. Fabrication of disconnected three-dimensional silver nanostructures in a polymer matrix. Appl Phys Lett. 2012;100(6):063120.

16. Okamuro K, Hashida M, Miyasaka Y, et al. Laser fluence dependence of periodic grating structures formed on metal surfaces under femtosecond laser pulse irradiation. Phys Rev B. 2010;82(16):165417.
17. Sakabe S, Hashida M, Tokita S, et al. Mechanism for self-formation of periodic grating structures on a metal surface by a femtosecond laser pulse. Phys Rev B. 2009;79(3):033409.
18. Wang J, Guo C. Ultrafast dynamics of femtosecond laser-induced periodic surface pattern formation on metals. Appl Phys Lett. 2005;87(25):251914.
19. Oliveira V, Ausset S, Vilar R. Surface micro/nanostructuring of titanium under stationary and non-stationary femtosecond laser irradiation. Appl Surf Sci. 2009;255(17):7556–60.
20. Nayak BK, Gupta MC, Kolasinski KW. Formation of nano-textured conical microstructures in titanium metal surface by femtosecond laser irradiation. Appl Phys A Mater Sci Process. 2008;90(3):399–402.
21. Amoruso S, Bruzzese R, Wang X, et al. Femtosecond laser ablation of nickel in vacuum. J Phys D: Appl Phys. 2007;40(2):331.
22. Ali N, Bashir S, Akram M, et al. Effect of dry and wet ambient environment on the pulsed laser ablation of titanium. Appl Surf Sci. 2013;270:49–57.
23. Khorsand S, Raeissi K, Ashrafizadeh F. Corrosion resistance and long-term durability of super-hydrophobic nickel film prepared by electrodeposition process. Appl Surf Sci. 2014;305:498–505.
24. Yang GW. Laser ablation in liquids: applications in the synthesis of nanocrystals. Prog Mater Sci. 2007;52(4):648–98.
25. Shaheen ME, Gagnon JE, Fryer BJ. Femtosecond laser ablation of brass in air and liquid media. J Appl Phys. 2013;113(21):213106.
26. Barmina EV, Stratakis E, Barberoglou M, et al. Laser-assisted nanostructuring of Tungsten in liquid environment. Appl Surf Sci. 2012;258(15):5898–902.
27. Lungu CP, Ticoş CM, Poroşnicu C, et al. Periodic striations on beryllium and tungsten surfaces by indirect femtosecond laser irradiation. Appl Phys Lett. 2014;104(10):101604.
28. Le Harzic R, Dörr D, Sauer D, et al. Large-area, uniform, high-spatial-frequency ripples generated on silicon using a nanojoule-femtosecond laser at high repetition rate. Opt Lett. 2011;36(2):229–31.
29. Vorobyev AY, Makin VS, Guo C. Brighter light sources from black metal: significant increase in emission efficiency of incandescent light sources. Phys Rev Lett. 2009;102(23):234301.
30. Søndergaard T, Novikov SM, Holmgaard T, et al. Plasmonic black gold by adiabatic nanofocusing and absorption of light in ultra-sharp convex grooves. Nat Commun. 2012;3:969.
31. Vorobyev AY, Topkov AN, Gurin OV, et al. Enhanced absorption of metals over ultrabroad electromagnetic spectrum. Appl Phys Lett. 2009;95(12):121106.
32. Amoruso S, Ausanio G, Barone AC, et al. Nanoparticles size modifications during femtosecond laser ablation of nickel in vacuum. Appl Surf Sci. 2007;254(4):1012–6.
33. Boinovich LB, Emelyanenko AM, Pashinin AS, et al. Origins of thermodynamically stable superhydrophobicity of boron nitride nanotubes coatings. Langmuir. 2011;28(2):1206–16.
34. Yong J, Chen F, Yang Q, et al. Femtosecond laser weaving superhydrophobic patterned PDMS surfaces with tunable adhesion. J Phys Chem C. 2013;117(47):24907–12.
35. Yong J, Yang Q, Chen F, et al. Superhydrophobic PDMS surfaces with three-dimensional (3D) pattern-dependent controllable adhesion. Appl Surf Sci. 2014;288:579–83.
36. Yong J, Yang Q, Chen F, et al. A simple way to achieve superhydrophobicity, controllable water adhesion, anisotropic sliding, and anisotropic wetting based on femtosecond-laser-induced line-patterned surfaces. J Mater Chem A. 2014;2(15):5499–507.
37. Yong J, Yang Q, Chen F, et al. Bioinspired superhydrophobic surfaces with directional Adhesion. RSC Adv. 2014;4(16):8138–43.
38. Li G, Li J, Zhang C, et al. Large-area one-step assembly of three-dimensional porous metal micro/nanocages by ethanol-assisted femtosecond laser irradiation for enhanced antireflection and hydrophobicity[J]. ACS applied materials and interfaces. 2014;7(1):383–390.

Chapter 6
Microcone Arrays by Sucrose Solution Assisted Femtosecond Laser Irradiation

After billions of years of evolution, fishes have developed perfect underwater superoleophobic ability, which makes their bodies well protected from oil contamination in aqueous environments, exhibiting appealing properties such as self-cleaning and antifouling [1, 2]. It is reported that fish scales consist of superhydrophilic micropapillae covered with nanostructures, exhibiting superhydrophilic in air. Hence, water can be trapped in the hierarchical micro/nanostructure to form an oil repellent layer, leading to underwater superoleophobicity and ultralow oil-adhesion [1, 2]. Inspired by fish scales, many kinds of underwater superoleophobic and low-adhesive interfaces have been fabricated with diverse technologies to boost their applications in self-cleaning coatings, fluidic devices, oil/water separation, bioadhesion, oil-droplet manipulation and so on [1–16].

Here, we demonstrate a facile and rapid method to fabricate large-area microcone arrays by one-step femtosecond laser irradiation on nickel targets in sucrose solutions. The prepared surfaces are superhydrophilic in air and superoleophobic in water. It is indicated that the height and size are tunable by controlling the ratio of sucrose to water in solution and the laser pulse energy, leading to controllable wettabilities for water and oil. Furthermore, these processed surfaces covered with microcone arrays exhibit multiple functions, which can be used in microfluidic devices, fluid microreactors, biomedicine, biomedical scaffolds, and chemical and biological sensors [18–23].

6.1 Experimental Section

Nickel sheets with a purity of 99.99% were purchased from New Metal Material Tech. Co., Ltd, Beijing, China. The oil used in our experiment is 1, 2-dichloroethane, and its chemical formula is $C_2H_4Cl_2$. Sucrose (chemical formula is $C_{12}H_{22}O_{11}$ and the relative molecular mass is 342.3), milk (the color is white) and red ink are purchased from a supermarket. The sucrose solution is prepared by dissolving a certain mass of sucrose in 100 g distilled water, and the mass ratio ranges from 10:100 to 70:100.

G. Li, *Bionic Functional Structures by Femtosecond Laser Micro/nanofabrication Technologies*, Springer Theses, https://doi.org/10.1007/978-981-13-0359-3_6

To structure the nickel surface, the incident laser pulse energies are set from 0.10 to 0.20 mJ per pulse, the scanning speed is 1 mm/s, and the step size between the adjacent scanning lines is 50 μm.

6.2 Preparation of Microcone Arrays

Figure 6.1a shows that the nickel surface irradiated in the sucrose solution with a mass ratio of 55:100 for sucrose and water and a laser pulse of 0.19 mJ displays a gray color. It is found that the water rapidly spreads out once it contacts with the structured surface. The measured contact angle is less than 5°, exhibiting superhydrophilicity. Furthermore, the oil droplet can spread out much faster than water and forms a smaller contact angle due to the smaller surface tension of oil than that of water, displaying superoleophilicity in air. In contrast, the processed nickel surface shows underwater superoleophobicity with an oil contact angle of nearly 170°.

Previously, we have reported the assembly of self-organized 3D porous nickel micro/nanocage arrays by ethanol assisted femtosecond laser irradiation. In direct contrast to the irradiation in sucrose solution, the processed nickel surfaces show a black color, and the surface is hydrophobic and oleophilic in air (Fig. 6.1d). The formation of the hydrophobic and oleophilic surface is related to the laser induced surface rough porous micro/nanocage structures (Fig. 6.1e) and the chemical composition [5–11]. The XRD analysis verified NiO was formed after the ethanol-assisted femtosecond laser irradiation (Fig. 6.1f), and the chemical formula can be described as

$$Ni + C_2H_5OH \rightarrow NiO \tag{6.1}$$

The micro/nanostructures formed in the two liquids are very different. In sucrose solution, hierarchical cone arrays with a size (here, we define the size of the cone as the cone bottom radius, which is labeled as "*R*") of 3 μm covered by 10–30 nm nanovillus structures are induced (Fig. 6.1b). In addition, it is almost difficult to see the flame and there appear only a few bubbles during the course of irradiation. The processing in sucrose solution is peaceful. These distinct features promote the formation of hierarchical cone arrays (Fig. 6.2) [23]. The nickel with a clean surface is initially hydrophilic. In addition, the XRD study shows that there is no chemical change on the irradiated nickel, so the produced pure nickel roughness enhances the hydrophilicity in air (Fig. 6.3) [23]. Figure 6.3a, b denote XRD curves of nickel processed in mass ratio of 10:100, under pulse energy of 0.10 and 0.19 mJ, respectively. Figure 6.3c indicates that the XRD results as a function of mass ratio of 55:100, pulse energy of 0.10 mJ. It is observed that the peaks appear at 44.5°, 51.8°, and 76.4°, which mean the positions for Ni. From the XRD curves, there is no other peak seen. Combined with the XRD analysis, the measurement covers the minimal and maximal mass ratio and pulse energy, which can certify powerfully that no chemical changes happen in sucrose solution.

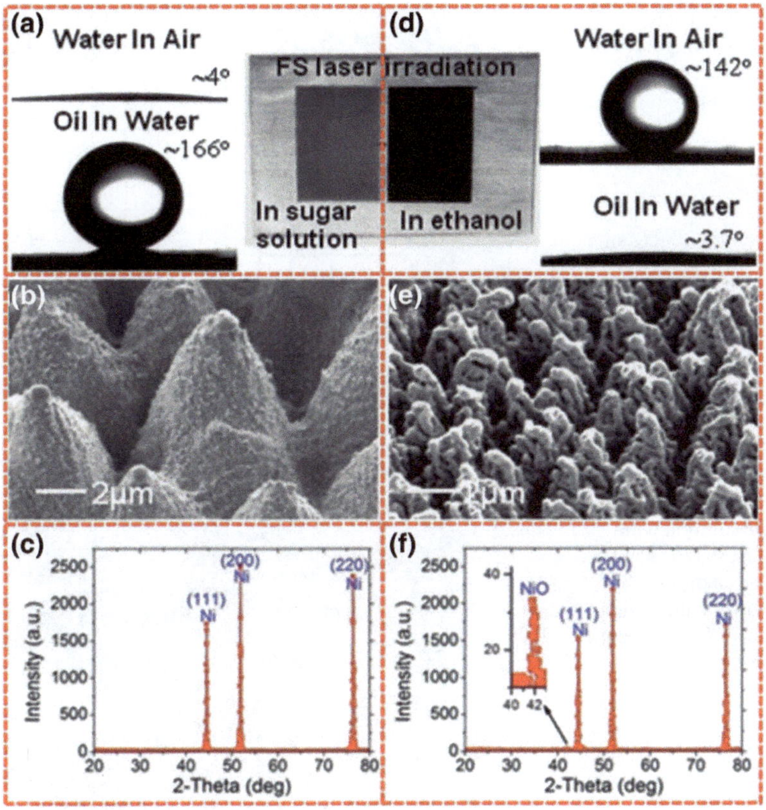

Fig. 6.1 The preparation of microcones by sucrose solution-assisted femtosecond laser irradiation

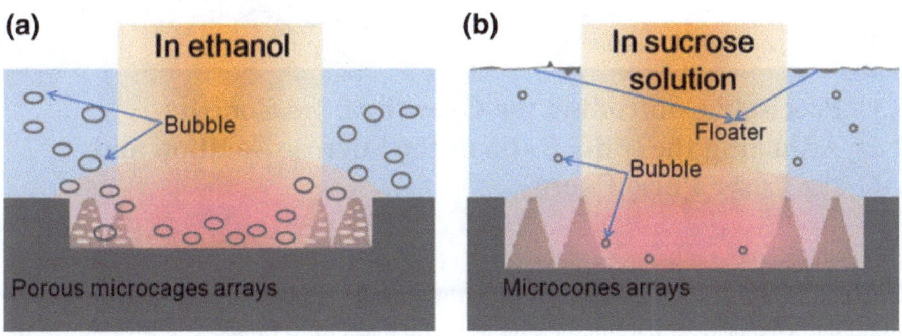

Fig. 6.2 Comparison of experimental phenomena for preparing porous microcages arrays in ethanol and microcones arrays in sucrose solution

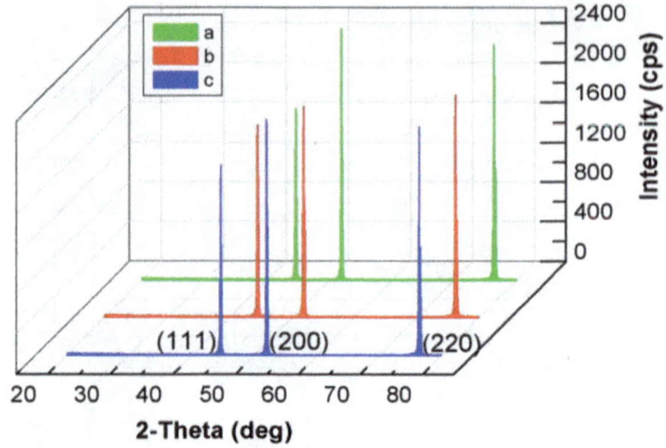

Fig. 6.3 XRD analysis of the chemical component of the nickel processed in sucrose solution

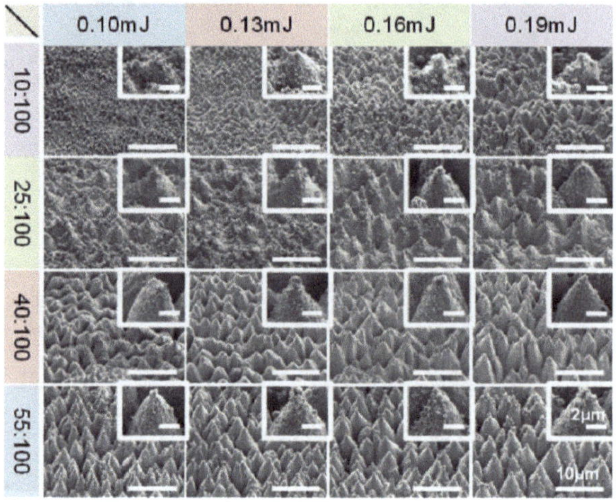

Fig. 6.4 45° tilted SEM images of the samples by femtosecond laser irradiation in sucrose solution with different mass ratios and laser pulse energies. The insets show the magnified SEM images for microcones, and the bar is 2 mm

6.3 Precisely Controlling the Growth of the Cone by Adjusting the Mass Ratio of the Sucrose Solution and the Laser Pulse Energy

From the aspect of practical application, it is acceptable to fabricate the controllable structures via a simple parameter. According to the experimental observation, it is found that two critical parameters, namely the mass ratio of sucrose to water in the solution and laser pulse energy should be carefully controlled. In experiments, four typical mass ratios ranging from 10:100 to 55:100 with pulse energies ranging from 0.10 to 0.19 mJ were examined (Fig. 6.4) [23].

It is easily seen that the mass ratio of the solutions greatly affected the cone topography. With the increasing mass ratio, the refractive index and viscosity of the solution can be sharply increased, leading to the stronger penetration into the

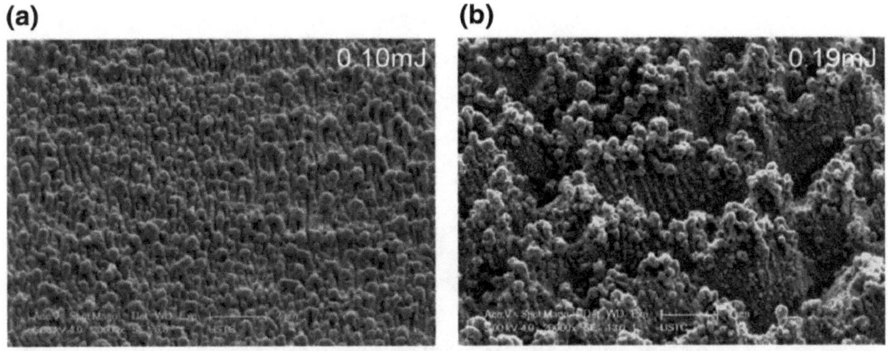

Fig. 6.5 SEM images of nickel surfaces processed in sucrose solution with mass ratio of 5:100. **a** The surface treated under pulse energy of 0.10 mJ; **b** The surface prepared under 0.19 mJ

nickel sheet, which plays an obvious role in the growth of the height and size. For example, when the mass ratio is increased from 10:100 to 55:100, the height and the size can increase from 1.62 to 8.02 μm and 2.1 to 2.63 μm at pulse energy of 0.10 mJ, respectively. In addition, the laser pulse energy was also used to control cones. As the pulse energy increases, the cones become larger due to the stronger interaction of laser with nickel surfaces. Taking a mass ratio of 10:100 for example, at low pulse energy of 0.10 mJ, the laser can only induce a small cone with a height of 2.14 μm and a size of 2.31 μm. With the increase of pulse energy, the laser-induced plasma is further expanded to produce a stronger shock wave, higher temperature, and higher pressure at the interface, so higher and larger cones are produced. By carefully modulating these two parameters, a series of representative microcone arrays are prepared (Fig. 6.4) [23]. It is noteworthy that the growth rate of the cone height caused by the mass ratio is faster than that of the pulse energy. The possible reasons for the growth of the cones are the higher boiling point and higher viscosity of the sucrose solutions with greater mass ratio.

Particularly, the experiments of sucrose solution with mass ratio of 5:100 and 75:100 are conducted, and the results are shown in Figs. 6.5 and 6.6, respectively [23]. For mass ratio of 5:100, it can be seen that the surface is covered by particles with size from tens to hundreds nanometers, and no cone was formed at pulse energy of 0.10 mJ. However, at pulse energy of 0.19 mJ, there is still no cone generated, but only uneven ravine-like structures covered by nanoparticles formed on the processing area. This results show that under the mass ratio of 10:100, there is no cone structures can be prepared.

In the case of mass ratio of 75:100, although few irregular cones are formed at pulse energy of 0.10 mJ, there are plenty of various nanoparticles generated around the formed cones, and covered them completely. At pulse energy of 0.19 mJ, it is indicated that the cones appear on the whole processing area. However, not only the height and size are decreased compared with the good cones mentioned in the paper, but also they are very heterogeneous, showing weak controllability. In addition, the

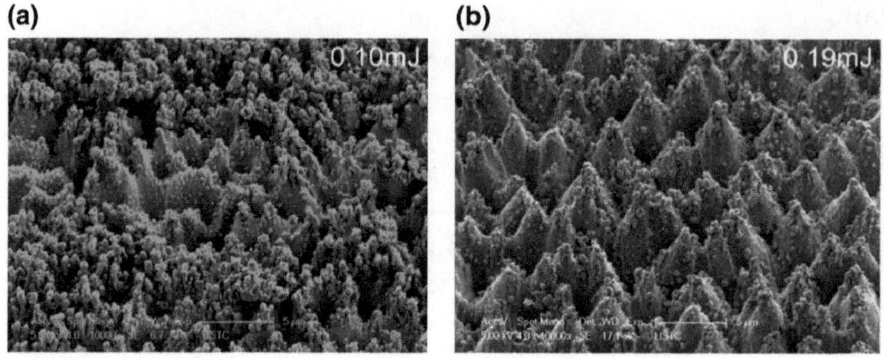

Fig. 6.6 SEM images of nickel surfaces processed in sucrose solution with mass ratio of 75:100. **a** The surface morphology of nickel prepared under pulse energy of 0.10 mJ; **b** The cones formed under the pulse energy of 0.19 mJ

top of the cones are covered with several to tens of nanoparticles. The formation mechanism can be attributed to the weak laser penetration through the high mass ratio sucrose solution.

The laser induced cone arrays are schematically illustrated in Fig. 6.7a. The microcone consists of the standard cone and the ambient villus nanostructures, which endows the microcones with hierarchical structures. The height and size for a single cone can be standardized as in Fig. 6.7b. In order to further investigate the influence of the mass ratio and pulse energy on the growth of the cones, the heights and the sizes as a function of pulse energy under different mass ratios are indicated in the histogram statistics (Figs. 6.7c, d), respectively. The statistical results demonstrated that the height and the size can be effectively modulated in the ranges of 2.1–2.81 μm and 1.6–10.34 μm, respectively. Compared with other methods reported before, our proposed method featured simplicity and flexibility.

6.4 The Wettabilities of the Microcone Arrays

The water contact angle of the flat nickel surface in our experiment is $\theta_{WA} = 58.26°$ (Fig. 6.8a), showing intrinsically weak hydrophilicity in air. Its wetting property can be tuned to be more hydrophilic and even superhydrophilic by the laser-induced rough hierarchical micro/nanostructures.

The roughness factor based on the data in Fig. 6.7c, d is calculated (Fig. 6.9) [23]. The roughness factor is defined as the ratio of the actual apparent surface area (S_1, the sector) to the geometric area (S_2, the circular). It is indicated that the roughness factor increased with the increasing microcone height and size. In addition, the nanoscale structures on the microcone also increase the superficial area of micro/nanocages, which further enhances the roughness factor. By introducing the roughness, the water

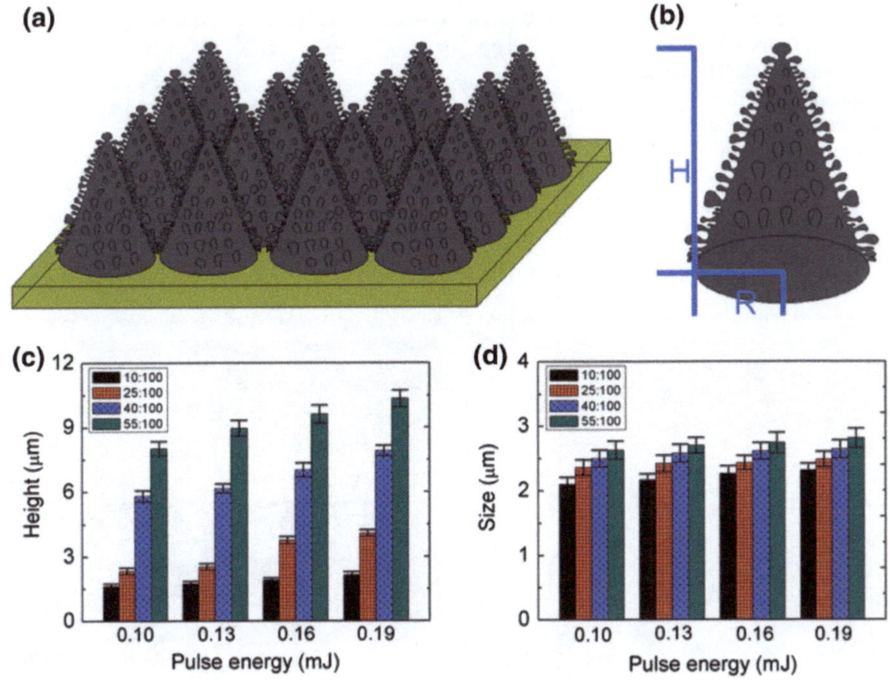

Fig. 6.7 **a** Schematic diagram for the prepared microcone arrays; **b** The schematic diagram of a single magnified microcone. The height (*H*) and the size (*R*) of the microcone are marked in the figure. **c** and **d** The statistical height and size of the microcones, which are obtained by calculating the average value

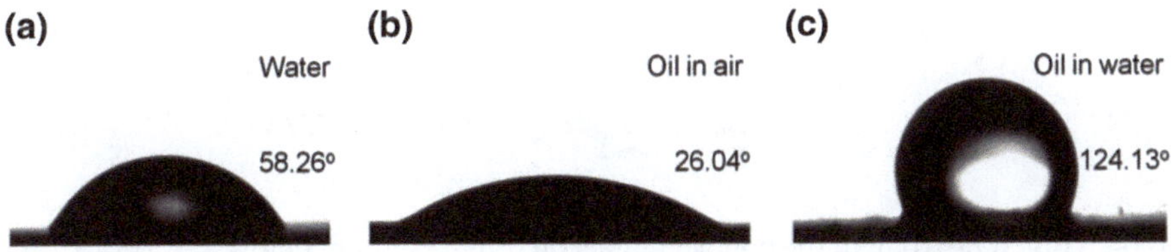

Fig. 6.8 **a** The water contact state in air. **b** The oil contact state in air. **c** The oil contact model underwater

contact angle of the nickel surface is modified to less than 14° (Fig. 6.10) [23]. Due to the hydrophilicity, the water can completely contact the rough surface, as shown in the schematic diagram of Fig. 6.10 [23]. The contact model for this case is described by the Wenzel mode.

According to Wenzel's model, the hydrophilic surfaces will be more hydrophilic with the increase of roughness. In addition, it can be observed that most surfaces are modified to be superhydrophilic with a contact angle less than 10°, and the minimum value is as small as 4°. On the other hand, the contact angle for oil on flat nickel is

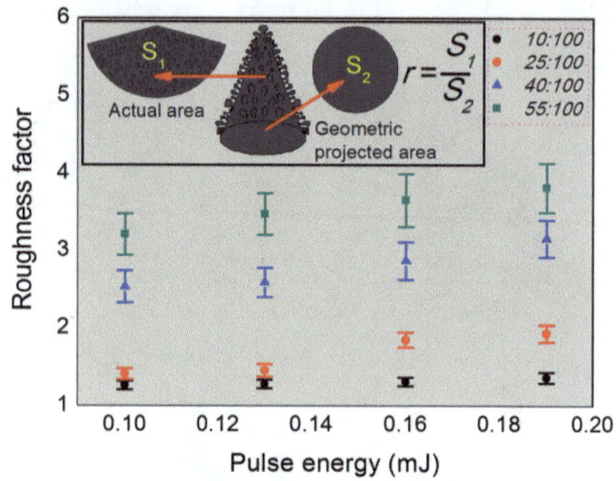

Fig. 6.9 The calculated roughness factor for the microcone arrays. The insets are the model of the microcone and the calculation formula

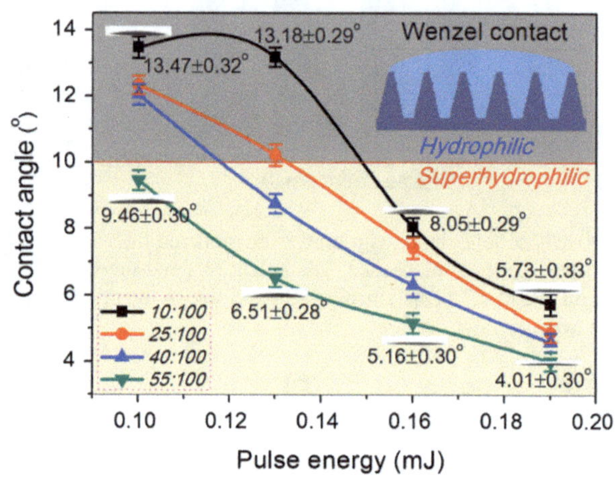

Fig. 6.10 The water contact angles

26.04° (Fig. 6.8b), but all the contact angles on rough ones are below 7.5° (Fig. 6.11), much less than that of water, demonstrating superoleophilicity [12–18, 23]. This is because of the lower surface tension of oil than that of water, which enables the oil droplet to be more easily spread on the rough nickel surface. Similarly, the oil contact angle is decreased with the increasing roughness, and the minimum value reached 3.5°.

It is indicated that the hydrophilic flat nickel surface is tuned to be an oleophobic one when it is immersed in water. Deduced from Young's equation, the oil contact angle can be described as the following equation:

$$\cos\theta_{OW} = \frac{\gamma_{OA}\cos\theta_{OA} - \gamma_{WA}\cos\theta_{WA}}{\gamma_{OW}} \tag{6.1}$$

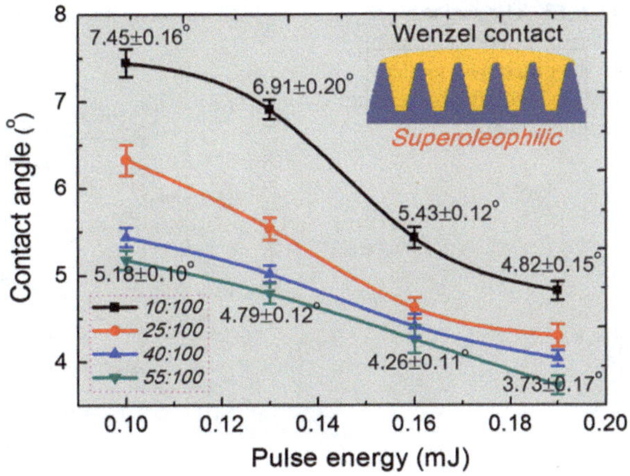

Fig. 6.11 The oil contact angles

where θ_{OW}, θ_{OA}, and θ_{WA} are the contact angle of oil on a flat surface in water, in air, and water contact angle on flat surface. γ_{OW}, γ_{OA}, and γ_{WA} are the surface tensions of the oil/water, oil/air, and water/air interfaces. For 1, 2-dichloroethane, the surface tension with air and water are 24.15 and 28.1 mN m^{-1}, respectively. The water surface tension is 73 mN m^{-1} at room temperature. Calculated from Eq. 6.4, the oil contact angle on the flat nickel surface in water is 126.47°, which is in good agreement with the measured value of 124.13° (Fig. 6.8c).

In Fig. 6.12, it is found that the underwater oleophobic flat nickel surface is tuned to be more oleophobic (135.22° < θ_{OW} < 150°) and even superoleophobic (θ_{OW} > 150°) by introducing the microcones. When the nickel is immersed in water, the processed area can be entirely wetted by water due to the superhydrophilicity. The water molecules can be trapped in the micro/nanostructures when the oil drop comes into contact with the processed surface, developing a composite water/solid interface. For the incompatibility, the trapped water acts as an insulation layer to prevent the oil droplet from permeating into the micro/nanostructures. In this case, the oil droplet can only contact the top of the induced micro/nanostructures, displaying oleophobic or superoleophobic surfaces in water. The state of the oil droplet can be described by the Cassie model, and the contact angle can be expressed as:

$$\cos\theta'_{OW} = f\,\cos\theta_{OW} + f - 1 \tag{6.2}$$

where θ'_{OW} is the contact angle of oil on the rough surface in water. f is the area fraction of the solid, which is defined as the ratio of the actual contact area by the oil droplet to the whole area of the cone. Although the exact value of f is not readily available, it can be theoretically deduced from Eq. 6.2, as shown in Fig. 6.13. In our study, the area fraction f is below 0.66, and the value decreases when the roughness increases. According to Eq. 6.2, for a rough nickel surface, the contact angle for underwater oil droplet is increased with decreasing area fraction.

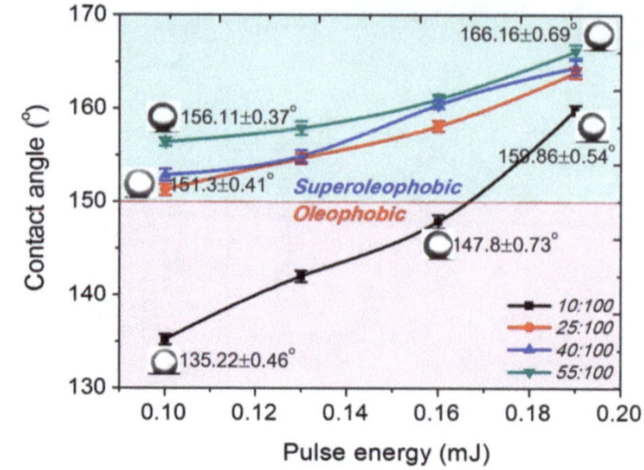

Fig. 6.12 The surfaces are underwater superoleophobic and the oil contact angles increased with the increasing roughness factor

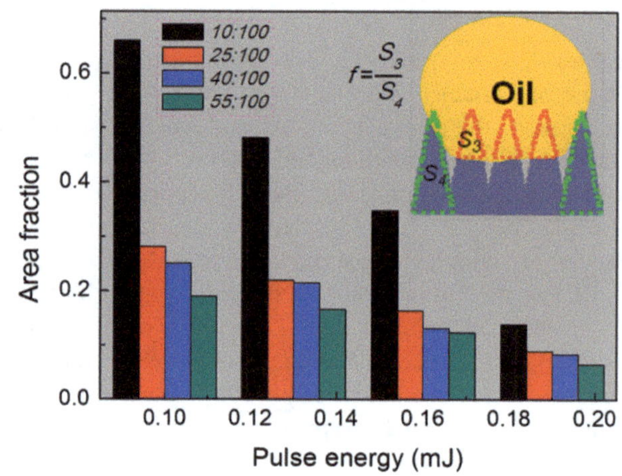

Fig. 6.13 The calculated area fraction which is defined as the ratio of actual contact area (S_3) by the oil droplet to the whole area of the cone (S_4)

It is observed from Fig. 6.12 that the measured contact angles on some surfaces with smaller roughness are below 150° [23]. According to the reported studies, these structured surfaces are in metastable Wenzel/Cassie hybrid states as shown in the schematic diagram of Fig. 6.14a. It is stated that some part of the oil droplet can homogeneously wet the rough surface, while the other part shows heterogeneous wetting, which can be more clearly observed from the magnified illustration in Fig. 6.14b. When the contact angle is above 150°, the contact is the Cassie model. In this case, the state represents a non-wet-contact mode, where the oil contacts the upper end of the microcones, leaving the lower end immersed in water, as shown in Fig. 6.14c. Especially, as indicated in Fig. 6.14d, when the area fraction f is about 0.066, the contact angle can nearly reach 170°, in which the oil droplet only contacts the peak of the cones, demonstrating extreme underwater superoleophobicity. Based on the above discussion, our prepared microcone arrays have the key properties of superhydrophilicity, superoleophilicity, and underwater superoleophobicity, which

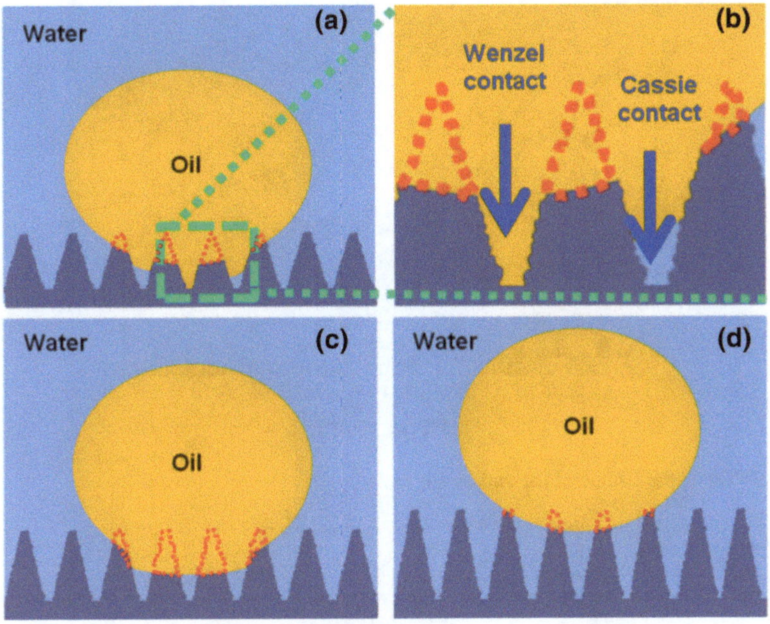

Fig. 6.14 The detailed contact model for underwater oil

are very similar to the features of the fish scales. Furthermore, it is indicated that all these wettabilities can be easily tuned over a wide range by adjusting the mass ratio and pulse energy.

6.5 The Liquid Manipulation

6.5.1 Liquid Droplet Transferring

It is reported that the oil can easily roll off the fish scales, keeping the fish body clean in contaminated water. The structured nickel surfaces have similar self-cleaning effects to fishes, which can be verified by measuring the oil sliding angle (SA) (Fig. 6.15). We found that the surfaces with different roughnesses showed different oil adhesions and SAs [19, 20]. The adhesion difference is ascribed to the distinct contact models. In the metastable state (Fig. 6.14a, b), the oil droplet partially wets the microcones and the oil sliding angle is higher than 9°. According to our previous discussion, the metastable states should appear in the region where the $CA_{OW} < 150°$, marked above the dashed line in Fig. 6.15 [23]. In contrast, below the dashed line, the contact modes belong to Cassie's state. The oil droplet can slide on the surface at a smaller sliding angle with lower adhesion. Especially, at a CA_{OW} nearly 170°, there is hardly any

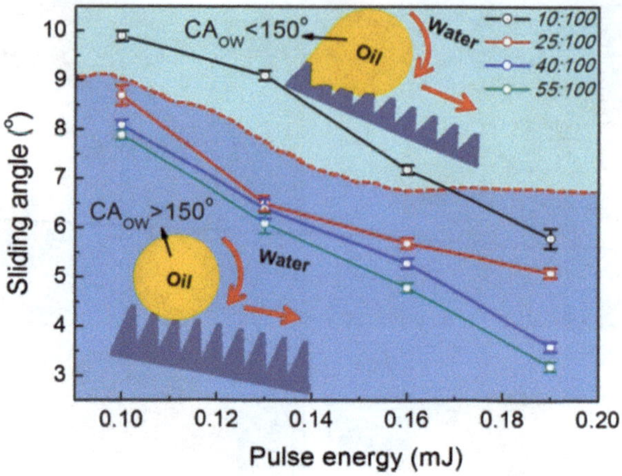

Fig. 6.15 The sliding angle of the underwater oil

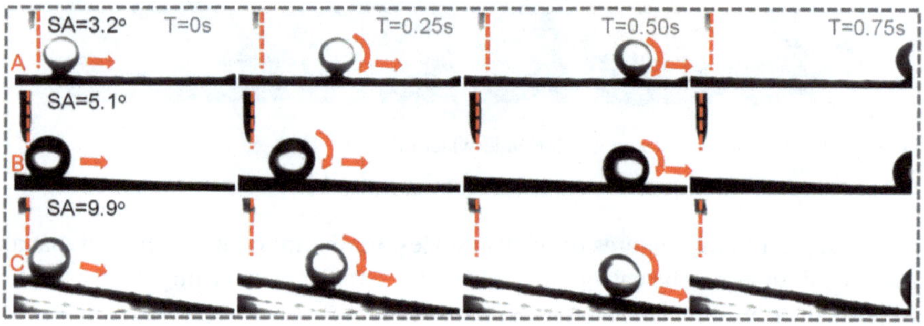

Fig. 6.16 The time sequences of snapshots of oil rolling on the three kinds of structured surfaces

contact space between the oil droplet and the rough structures, so that the adhesion is extremely low and the oil droplet easily rolls off with the surface slightly tilted at only 2°. This study indicates the adhesion force can be controllably tuned from the metastable state with high adhesion to the Cassie state with ultralow adhesion.

The sliding behaviors of three structured surfaces named "*A*", "*B*", and "*C*" are studied in detail (Fig. 6.16) [23]. Figure 6.16 shows the time sequences of snapshots of a 5 μl oil droplet rolling on the three surfaces. It can be seen that the oil droplet can move easily when the *A* surface is only slightly tilted 2.2°, and it rolls off the entire surface in a very short time. On the contrary, the oil can only slide a short distance on the *B* surface in the same time, and possesses a SA of 5.1°. As for the *C* surface, the oil droplet slides with an extremely slow speed even when the surface is tilted about 9.9°.

Another experiment about their adhesive behavior is conducted by controlling the oil droplet to contact and leave the *A*, *B*, and *C* surface (Fig. 6.17) [23]. For the *A* surface, during the whole course of contact and leave, the shape of the oil droplet

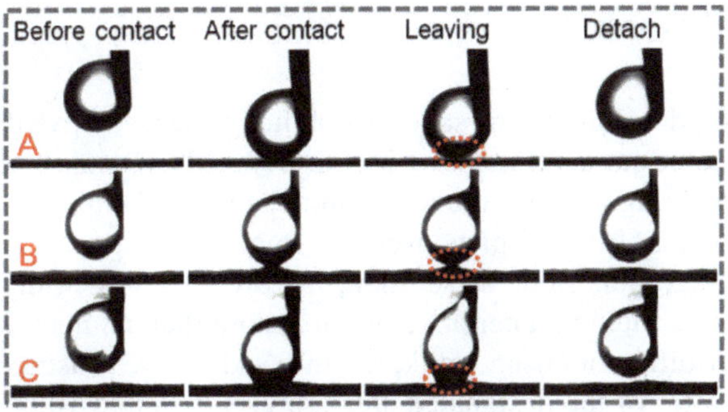

Fig. 6.17 Dynamical adhesive force behaviors on **A**, **B**, and **C** surfaces

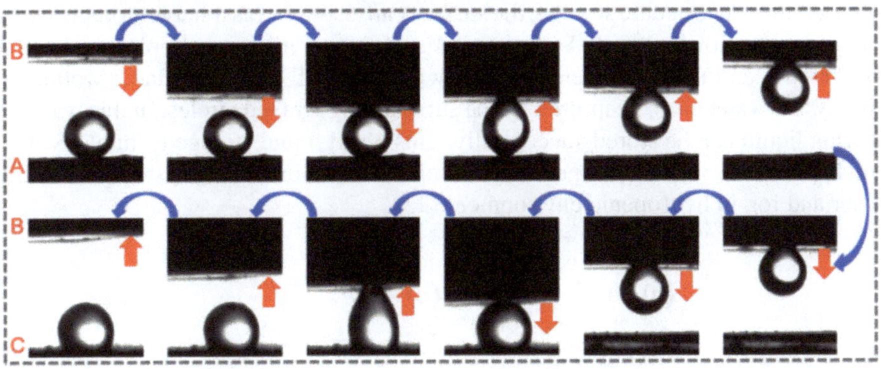

Fig. 6.18 Process of transferring the oil droplet from the **A** surface to the **C** surface via the **B** surface

remains spherical, with almost no changes, showing a very small adhesive force. In this case, the oil droplet can be transferred away without any loss. For the *B* surface, it is found that the oil droplet is first elongated to elliptical, and has a certain contact area with the *B* surface when it is about to leave. This phenomenon indicates that the *B* surface has larger adhesive force than that of the *A* surface. It is observed that the oil droplet is seriously deformed into a long elliptical shape and has a large contact area when it is pulled from the *C* surface, revealing strong adhesion.

These surfaces could be applied in the manipulation of oil droplets by transferring them from a lower adhesive surface to a higher one, which have key potential applications in biomedical fields. Figure 6.18 shows an example of transferring a 5 μl oil droplet. First, the oil droplet is placed at the low adhesive *A* surface. Then, the medium adhesive *B* surface is used as a "mechanical hand" to capture the oil droplet and release it onto the *C* surface [21, 23].

6.5.2 Liquid Storage

Due to the superhydrophilicity and superoleophilicity, the as-prepared nickel surfaces can be used for liquid storage. As shown in Fig. 6.19, milk, red ink, and oil are dripped onto the processed ring-shaped area with 5 mm diameter, forming 3 × 3 arrays [23]. It is found that none of the droplets overflow from the designated circular regions, displaying ultrahigh adhesion. In addition, the droplets still firmly adhere to the surface even if it is tilted at a certain angle for more than an hour. In addition, the static storage for different dosage milk, red ink, and oil are illustrated in Fig. 6.20, where the diameters of the designated area are 2, 3, 5, 8 and 10 mm, respectively [23]. With this method, a few microliters to several milliliters and an even larger dosage of various liquids can be stored at the designated area.

In addition to the static storage, the surface can also serve as a stable storage media for storing the flowing fluid. As shown in Fig. 6.21, a 5 ml water droplet can be first fixed at the central circle. Then the water will flow in all directions under wobbling. Finally, the water can be captured by the surrounding arrayed circles. In this way, the flowing liquid can be stored successfully. This novel liquid storage technology may find applications in fixing reagent in gas liquid reaction in sensing system, collecting water and fog in hydropenic environment [22].

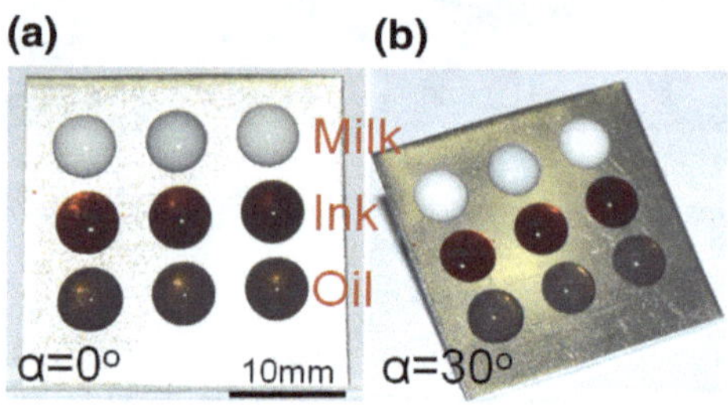

Fig. 6.19 The static storage for milk, ink and oil

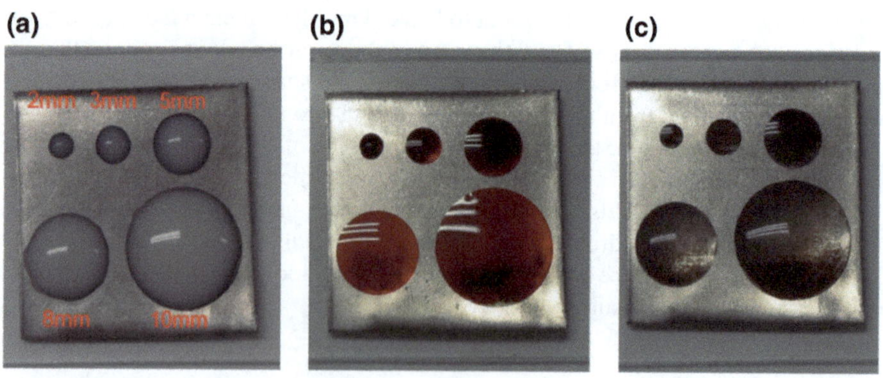

Fig. 6.20 The static storage for **a** milk, **b** red ink and **c** oil with different dosage. The diameters of the designated area are 2, 3, 5, 8 and 10 mm, respectively

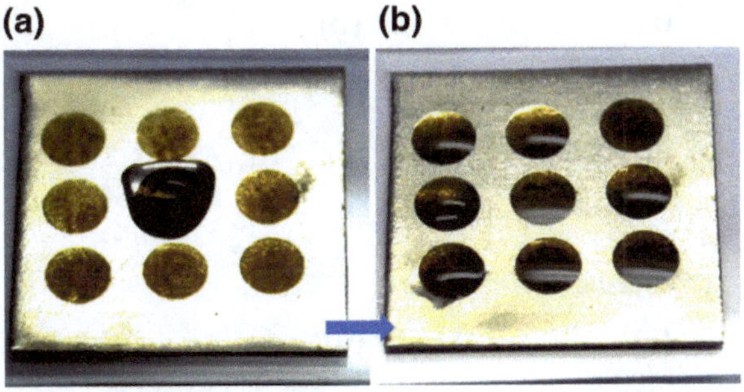

Fig. 6.21 The dynamic storage for water

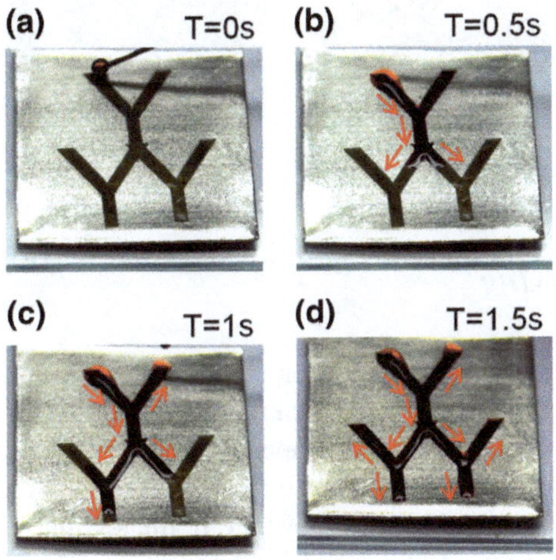

Fig. 6.22 The capillary force driving property for red ink

6.5.3 Liquid Transportation

Figure 6.22 shows the transportation dynamics of 5 ml red ink droplet pipetted on the processed surface. The track of the red ink is marked with a red dotted line in each figure. It is found that the red ink immediately flows along the set path once it contacts the starting point, and rapidly travels over 50 mm in less than 6 s. It is thought that the extremely strong driving force is derived from the wicking force due to the surface superhydrophilicity. This experiment indicates that a regular nickel surface can be readily transformed to a superwicking one, which may find potential applications in micro-fluidics, lab-on-chip, and so on [22, 23].

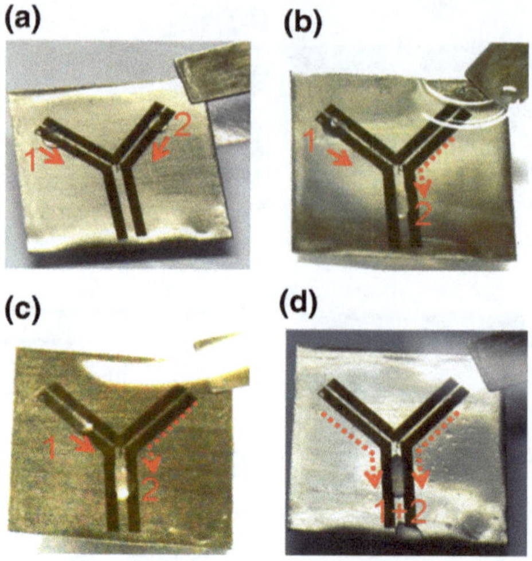

Fig. 6.23 The mixing function for oil in water

6.5.4 *Liquid Mixing*

Previous discussion indicates that in water, the untreated nickel is oleophobic, while the treated one is superoleophobic. By taking advantage of this property, we have designed a path to form a hollow "*Y*" shape, as shown in Fig. 6.23, composed of an untreated fine line and treated thick lines on both sides. In this way, the fine line can grasp the oil droplet and control it to move along the setting path, and the ultralow adhesion force of the processed lines on both sides ensures that the oil droplet moves without loss. As shown in Fig. 6.23, by gently vibrating the sheet, oil droplets "*1*" and "*2*" can travel along the "*Y*" shaped line and finally mix together [23].

In summary, inspired by the fish scales, hierarchical microcone arrays are one-step assembled on a nickel surface by sucrose solution-assisted femtosecond laser irradiation. The structured nickel surface is superhydrophilic, superoleophilic, and superoleophobic underwater, possessing similar properties to fish scales. In addition, by tuning the solvent environment and the pulse energy, the height and the size of the microcones can be controlled effectively. The controllable microcones endow the processed nickel surface with multifunctions, such as liquid transferring, storage, transportation and mixing. The unique wetting properties of the structured nickel open new applications in the fields of microfluidic devices, fluid microreactors, biomedicine, biomedical scaffolds, and chemical/biological sensors.

References

1. Liu M, Wang S, Wei Z, et al. Bioinspired design of a superoleophobic and low adhesive water/solid interface. Adv Mater. 2009;21(6):665–9.
2. Liu K, Tian Y, Jiang L. Bio-inspired superoleophobic and smart materials: design, fabrication, and application. Prog Mater Sci. 2013;58(4):503–64.
3. Xue Z, Jiang L. Bioinspired underwater superoleophobic surfaces. Acta Polym Sin. 2012;10:1091–101.
4. Liu X, Gao J, Xue Z, et al. Bioinspired oil strider floating at the oil/water interface supported by huge superoleophobic force. ACS Nano. 2012;6(6):5614–20.
5. Yang HD, Li XH, Li GQ, et al. Formation of colorized silicon by femtosecond laser pulses in different background gases. Appl Phys A: Mater Sci Process. 2011;104(2):749–53.
6. Amoruso S, Bruzzese R, Wang X, et al. Femtosecond laser ablation of nickel in vacuum. J Phys D: Appl Phys. 2007;40(2):331.
7. Ali N, Bashir S, Akram M, et al. Effect of dry and wet ambient environment on the pulsed laser ablation of titanium. Appl Surf Sci. 2013;270:49–57.
8. Bian H, Yang Q, Liu H, et al. A facile preparation route for netlike microstructures on a stainless steel using an ethanol-mediated femtosecond laser irradiation. Mater Sci Eng: C. 2013;33(2):663–7.
9. Barmina EV, Stratakis E, Fotakis K, et al. Generation of nanostructures on metals by laser ablation in liquids: new results. Quantum Electron. 2010;40(11):1012.
10. Kumar B, Yadav D, Thareja RK. Growth dynamics of nanoparticles in laser produced plasma in liquid ambient. J Appl Phys. 2011;110(7):074903.
11. Albu C, Dinescu A, Filipescu M, et al. Periodical structures induced by femtosecond laser on metals in air and liquid environments. Appl Surf Sci. 2013;278:347–51.
12. Brackbill JU, Kothe DB, Zemach C. A continuum method for modeling surface tension. J Comput Phys. 1992;100(2):335–54.
13. Tyson WR, Miller WA. Surface free energies of solid metals: Estimation from liquid surface tension measurements. Surf Sci. 1977;62(1):267–76.
14. Tolman RC. The effect of droplet size on surface tension. J Chem Phys. 1949;17(3):333–7.
15. Rotenberg Y, Boruvka L, Neumann AW. Determination of surface tension and contact angle from the shapes of axisymmetric fluid interfaces. J Colloid Interface Sci. 1983;93(1):169–83.
16. Kirkwood JG, Buff FP. The statistical mechanical theory of surface tension. J Chem Phys. 1949;17(3):338–43.
17. Van Oss CJ, Good RJ, Chaudhury MK. Additive and nonadditive surface tension components and the interpretation of contact angles. Langmuir. 1988;4(4):884–91.
18. Jasper JJ. The surface tension of pure liquid compounds. J Phys Chem Ref Data. 1972;1(4):841–1010.
19. Yong J, Chen F, Yang Q, et al. Femtosecond laser weaving superhydrophobic patterned PDMS surfaces with tunable adhesion. J Phys Chem C. 2013;117(47):24907–12.
20. Yong J, Yang Q, Chen F, et al. Reversible underwater lossless oil droplet transportation. Adv Mater Interfaces. 2015;2(2).
21. Yong J, Yang Q, Chen F, et al. A simple way to achieve superhydrophobicity, controllable water adhesion, anisotropic sliding, and anisotropic wetting based on femtosecond-laser-induced line-patterned surfaces. J Mater Chem A. 2014;2(15):5499–507.
22. Huang JY, Lai YK, Pan F, et al. Multifunctional superamphiphobic TiO_2 nanostructure surfaces with facile wettability and adhesion engineering. Small. 2014;10(23):4865–73.
23. Li G, Lu Y, Wu P, et al. Fish scale inspired design of underwater superoleophobic microcone arrays by sucrose solution assisted femtosecond laser irradiation for multifunctional liquid manipulation [J]. J Mater Chem A. 2015;3(36):18675–18683.

References

Chapter 7
Fabrication of Controllable Microcone and Micromolar Silicon Arrays with Tunable Wettability by Liquid-Assisted Femtosecond Laser Irradiation

In this work, we use low-boiling, low viscosity ethanol, and for the first time employ the high boiling point and high viscosity sucrose as liquid environment for femtosecond laser irradiation on silicon. The microcone and micromolar arrays were successfully prepared in ethanol and sucrose, respectively. It is indicated that the heights are tunable by simply controlling the laser pulse energy. Furthermore, these processed surfaces exhibit multiple functions, such as collecting fog, liquid storage, microdroplet manipulation and directional transfer, which can be used in microfluidic devices, fluid microreactors, biomedicine, biomedical scaffolds, and chemical/biological sensors [1–5].

7.1 Experimental Section

In this experiment, single crystal p-type silicon (110) with dimension of $0.5 \times 20 \times 20\ mm^3$ and $0.5 \times 20 \times 40\ mm^3$ were purchased from New Metal Material Tech. Co., Ltd, Beijing, China. The oil used to detect the contact/sliding angles is 1, 2-dichloroethane ($C_2H_4Cl_2$). The assisted solvents are ethanol (purity of 99.3%) and sucrose which is prepared by dissolving 50 g sucrose in 50 g distilled water.

The details of the experimental setup and schematic fabrication process are shown in Fig. 7.1a [5]. To induce micro/nanostructures with different heights at large area, the pulse energies of the incident laser are set from 0.05 to 0.25 mJ, and the scanning speed is constant (1 mm/s). The step sizes between the adjacent scanning lines are 20, 35, and 50 μm, respectively. For liquid experiments, the silicon sheet was stuck to the bottom of a 20 ml glass container. When the glass container was filled up with ethanol or sucrose, the height of the liquid layer above the sample surface was about 12 mm. In this experiment, the focusing condition is an vitally important parameter, which can greatly influence the energy distribution and morphology of the formed micro/nanostructures. In this experiment, the focusing condition is an vitally important parameter, which can greatly influence the energy distribution and morphology of the formed micro/nanostructures [5].

G. Li, *Bionic Functional Structures by Femtosecond Laser Micro/nanofabrication Technologies*, Springer Theses, https://doi.org/10.1007/978-981-13-0359-3_7

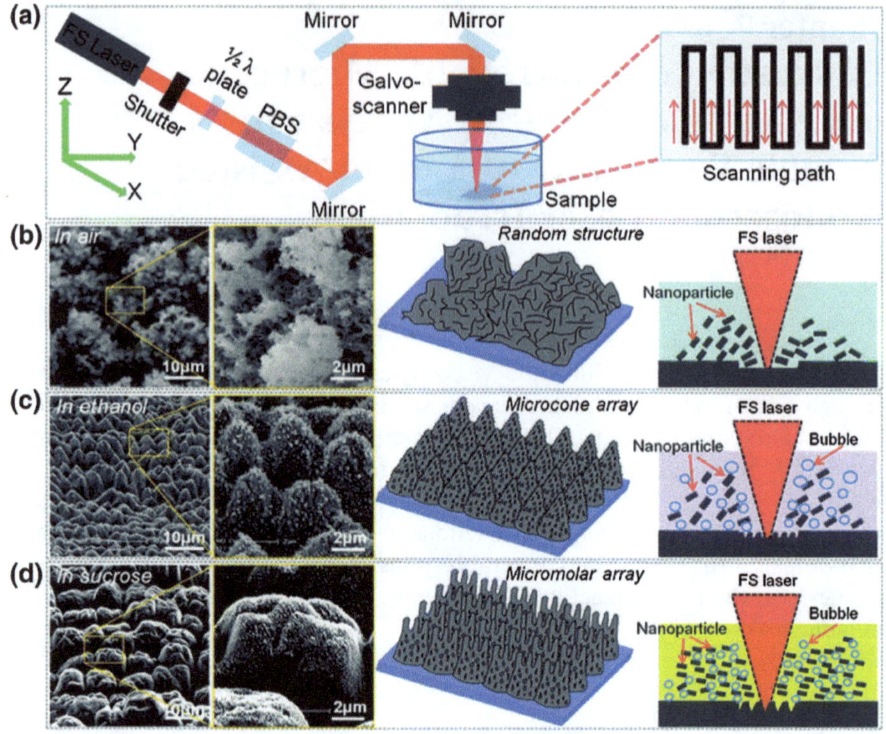

Fig. 7.1 The preparation of microcones and micromolars with ethanol and sucrose solution assisted fs laser irradiation at pulse energy of 0.06 mJ. **a** Schematic representation of the experimental setup. **b** Random structures in air under the same laser parameters. The structures consisted of cotton-like structures covered with debris of silicon. **c** Microcones arrays with ethanol assisted fs laser irradiation. The microcones height is about 3.3 μm. During the experiment, few but large bubbles were produced at the ethanol/silicon interface, and nanoparticles were also formed. **d** Micromolars arrays with sucrose solution assisted fs laser irradiation. The morphology of the microstuctures like mammal molars and the height is about 5.9 μm. A lot of tiny bubbles and nanoparticles were generated during the experiment

The glass container (vessel) containing the sample was mounted on a computer controlled *XYZ* translation stage for precisely positioning of the area to be machined. Considering the different focal conditions in air and liquids due to refractive effects, we adjusted the position of focal plane by moving the sample along the optical axis (*Z* direction) in order to obtain the same focal positions. During the adjustment process, a CCD camera was used for in situ observation. The detailed description of the focusing conditions is shown Fig. 7.2 [5].

Within the experimental setup of pulsed laser ablation in liquid, the laser beam has to penetrate the liquid layer before reaching the material surface. During this process, the focal length of the focusing lens will change due to refraction by the liquid layer. Considering the refraction of a focused linear beam, the focal length will increase [6, 7] for

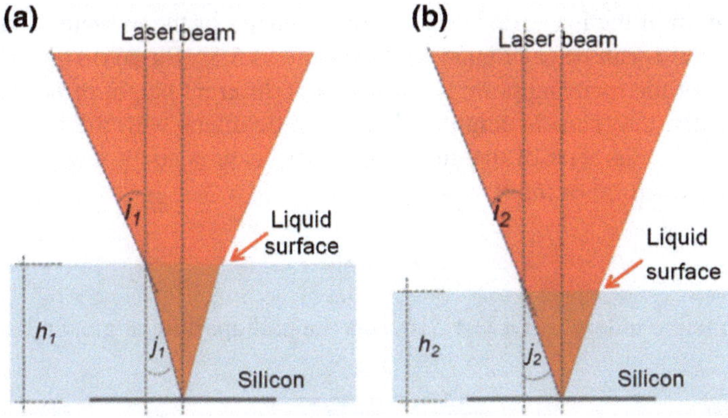

Fig. 7.2 The focus conditions for the laser beam in liquid with different height. **a** height of h_1, **b** height of h_2

$$\Delta f = l\left(1 - \frac{f}{\sqrt{n^2 f^2 + (n^2 - 1)r^2}}\right) \tag{7.1}$$

where f is the focal length of the focusing lens in air, l is the liquid thickness, n is the refractive index of the liquid, and r is the radius of unfocused laser beam. For $r \ll f$, the formula can be simplified to

$$\Delta f = l\left(1 - \frac{1}{n}\right) \tag{7.2}$$

Hence, the amount of energy deposited on the surface will vary for the varying focused position. From Eq. (7.2), it can be calculated that the focal length will increase 3.176 and 3.304 for ethanol ($n = 1.36$) and sucrose ($n = 1.38$), respectively. In order to obtain the same focusing condition, we move the sample surface away at length of Δf to achieve optimum focusing. Monitoring of the adjustment was accomplished by a CCD camera with a video monitor. It was indicated that the position of focal point was on the silicon surface, with the error ± 1 μm.

The focusing in different height of liquid layer can also be illustrated in Fig. 7.2. With the liquid height of h_1 (Fig. 7.2a) and h_2 (Fig. 7.2b), the refractive index can be expressed as:

$$n_1 = \frac{\sin i_1}{\sin j_1} \tag{7.3}$$

$$n_2 = \frac{\sin i_2}{\sin j_2} \tag{7.4}$$

For a certain liquid, the value of n_1 is equal to that of n_2 and the laser beam incident angle i_1 is equal to i_2, hence the refraction angle is equal, namely $j_1 = j_2$. Combing the focal length of the lens of 63 mm and the diameter of the unfocused laser beam of 4 mm, j_1 and j_2 can be calculated, and the value is 5.5°. Based on this discussion, the diameter of the focusing point is consistent at different height of liquid layer.

Additionally, the focus is completed by a flat field lens with the focal length of $d = 63$ mm. The diameter of the unfocused laser beam is about $r = 4$ mm. Hence, the numerical aperture of the lens is:

$$NA = n \sin \alpha \tag{7.5}$$

n is the refractive index, and $n = 1.516$. α is the half aperture angle, and

$$\sin \alpha = \frac{r}{\sqrt{d^2 + r^2}} \tag{7.6}$$

The diameter of the laser beam do can be expressed as

$$d_o = \frac{k\lambda}{NA} \tag{7.7}$$

here, k is a constant and the value is 1.22. λ is the laser wavelength and the value is 800 nm. Combining Eqs. (7.1)–(7.3), it can be calculated that $d_o = 13.5$ μm. The focus point is so small that the microstructures cannot be induced if the silicon is not at the focal plane.

7.2 Formation of Microcone and Micromolar Arrays in Ethanol and Sucrose Solutions

The ethanol has the high density of 99.3%, low boiling point of 78.4 °C and low viscosity. On the contrary, the sucrose solution has the high boiling point (higher than 100 °C) and high viscosity. The physical properties for the two solutions are very different. So, the laser interaction mechanism between silicon and different solvent may be different. Moreover, in our previous works, we have reported the one-step assembly of 3D porous metal micro/nanocages by ethanol-assisted femtosecond laser irradiation for enhanced antireflection and hydrophobicity, and the preparation of underwater superoleophobic microcones arrays by sucrose solution assisted femtosecond laser irradiation. In this work, we also used ethanol and sucrose solution as liquid environment to assist the femtosecond laser fabrication on silicon surfaces.

SEM images in Fig. 7.1b indicate the silicon surface micro/nanostructures prepared in air at pulse energy of 0.06 mJ, scanning speed of 1 mm s^{-1}, and scanning space of 50 μm. It is observed that the structures are mainly nanoparticles covered cloud-structures with various sizes, displaying random, which can be attributed to silicon vapor cloud formed on and above the surface during its heating by the laser radiation, and the redeposition of the multiscale nanoparticles produced by the laser impact effect. Figure 7.1c exhibits that the formation of the microcones with the ethanol-assisted fs laser irradiation. It is indicated that the microcones are regularly distributed and the average height is about 3.3 μm. Different from the micro/nanostructures prepared in ethanol, the formed micro/nanostructures in sucrose solution can be described as micromolars (Fig. 7.1d) with the height of 5.9 μm. It is seen that the microcones and micromolars are distributed into arrays [5].

The mechanism for the formation of such structures can be ascribed to the high temperature of plasma excited at the silicon/liquid interface and the capillary wave excited in the molten silicon thin layer during the localized cooling process. By contrast, it is observed that during the machining process, there were more and smaller bubbles produced than that in ethanol. In addition, the processing in sucrose solution is more peaceful. The differences of density, viscosity as well as boiling point of the ethanol and sucrose also have great influence in inducing the diverse microstructures on silicon surface. Therefore, the ambient conditions are critical for the formation of different micro/nanostructures.

7.3 Precisely Controlling the Growth of the Microcones and Micromolars Arrays by the Laser Pulse Energy

From the aspect of practical application, it is acceptable to fabricate controllable structures via a simple parameter. Based on the experimental observation, the growth of both the microcones and micromolars arrays can be effectively controlled by adjusting the laser pulse energy. In order to illustrate this controllability, a series of pulse energy ranged from 0.05 to 0.25 mJ is chosen to do the irradiation experiments. With the increasing pulse energy, the laser-induced plasma is further expanded to produce a stronger shock wave, higher temperature and higher pressure at the interface, so higher and larger microcones and micromolars are produced. For example, when the pulse energy is increased from 0.05 to 0.25 mJ, the height can be increased from 3.3 to 17.6 μm for microcones (Fig. 7.3a) and from 5.9 to 33.7 μm for micromolars (Fig. 7.3b), respectively (Fig. 7.4). Additionally, it can be seen that the grooves with the adjacent spaces of 50 μm are also formed between the micro/nanostructures, with the period equal to the direction translation step due to the scanning of the laser beam [5].

In order to further investigate the evolution of the microcones and micromolars, and the formation of the grooves, we conduct another experiment with the pulse energy ranged from 0.06 to 0.12 mJ, and the scanning spaces of 20 μm (Fig. 7.3c, d). In Fig. 7.3c, at pulse energy of 0.06 mJ, it is seen that small microcones with height of 3–3.5 μm are partly generated on the processed area. With the increasing pulse energy, the microcones gradually grow larger, and the height is 4.1–5.2 μm at 0.08 mJ, 5.5–6.3 μm at 0.10 mJ, and 6.5–7.1 μm at 0.12 mJ. In

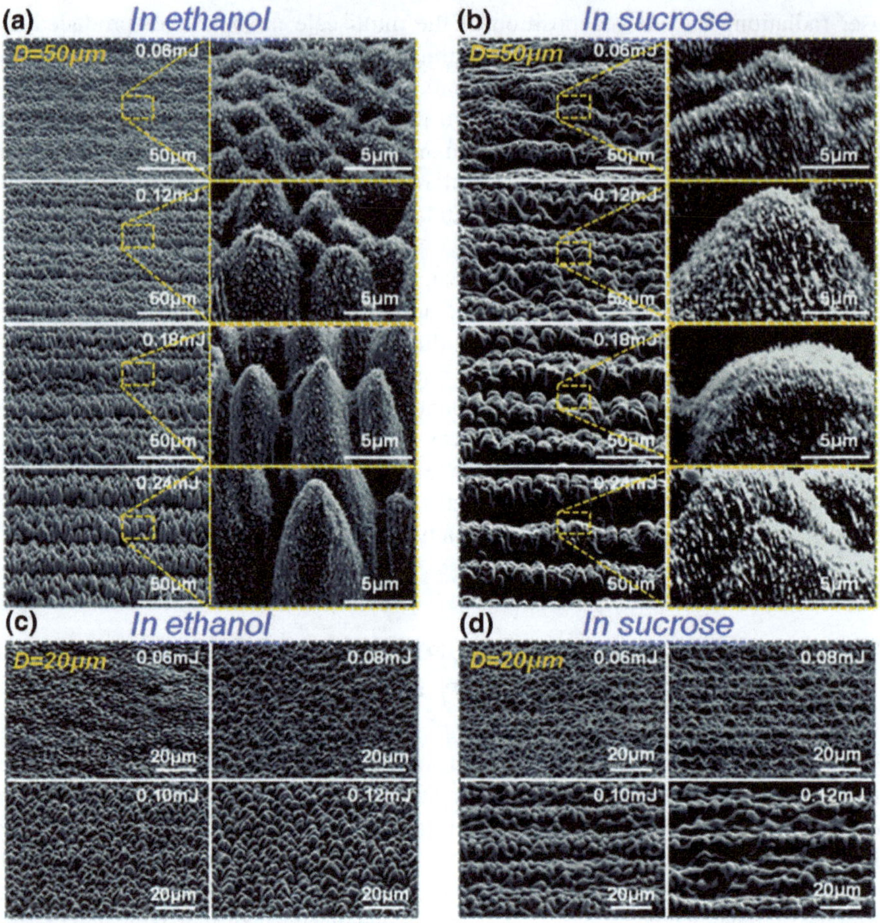

Fig. 7.3 Precisely controlling the growth of the microcone and micromolar arrays by adjusting the laser pulse energy. **a** 45° tilted SEM images of the silicon surfaces by fs laser irradiation in ethanol with D = 50 μm at pulse energy from 0.06 to 0.24 mJ. The microcones are uniformly distributed and the height is increased from 3.3 to 17.6 μm with the increasing pulse energy. **b** 45° tilted SEM images of the micromolars fabricated in sucrose solution. The corresponding height is ranged from 5.9 to 33.7 μm. **c** and **d** SEM images of the microcones and micromolars prepared at D = 20 μm, pulse energy of 0.06~0.12 mJ, and the energy interval is 0.02 mJ. Under this processing condition, all the microcones are still uniformly distributed on the whole processed surface, however for micromolars there are always groove caused by the scanning lines

addition, the distributions become more uniform with the increasing pulse energy. The SEM images in Fig. 7.3d indicate that the micromolars also grow as a function of the pulse energy, and the height can reach about 9.4 μm at 0.12 mJ. By comparing the results in Fig. 7.3c, d, it is found that the micromolars are always larger than the microcones produced at the same parameter. Another distinct difference is that the grooves can be always formed under the sucrose solution assisted fs laser irradia-

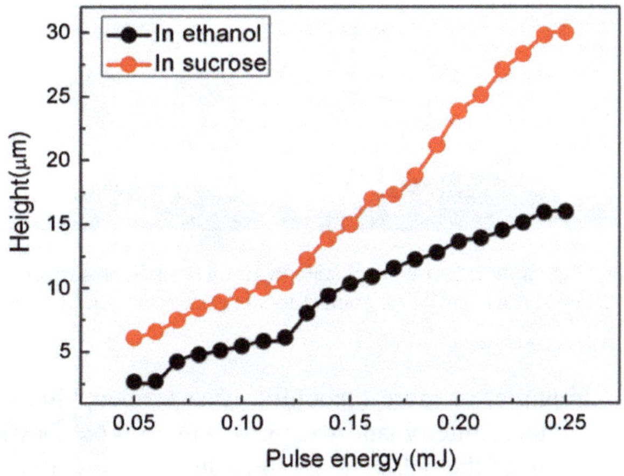

Fig. 7.4 The height of the microstructures as functions of pulse energy

tion. The generation mechanism for these differences can be attributed to the different boiling points, viscosity, and refractive index of ethanol and sucrose solution. Mazur [8] has reported the formation silicon spikes by liquid (water) assisted femtosecond laser pulses irradiation, however, the size of the structures are 200 nm, which are obviously smaller than laser wavelength of 800 nm. Chen et al. [1, 2] has presented an effective method of producing hierarchical conical spike structures femtosecond laser irradiation in air through a line-by-line and serial scanning process. It is indicated the formation of the conical spikes can be attributed to the femtosecond laser holing, and the period of the structures is equal to the spaces of the adjacent pulse, which is different from our strategy. This experiment demonstrates that a simple and flexible method have been proposed to fabricate controllable microcones and micromolars on silicon surfaces [5].

7.4 The Wettability of the Microcones and Micromolars Arrays

7.4.1 The Superhydrophilicity/Superoleophilicity

Mazur [9] has already investigated the wettability of the structured silicon surface, however, they focused on the superhydrophobicity which is obtained by modified the surface with a layer of fluoroalkylsilane molecules. Here, different from the literature, we mainly studied the superhydrophilicity or superoleophilicity of the micro/nanostructured surface. In our experiment, the water and oil contact angle of the flat silicon surface is about 36.2° and 24.7° in air, demonstrating intrinsically weak hydrophilicity and oleophilicity, respectively (Fig. 7.5) [5]. It is reported that the hydrophilic/oleophilic flat surface can be tuned to be more hydrophilic or super-

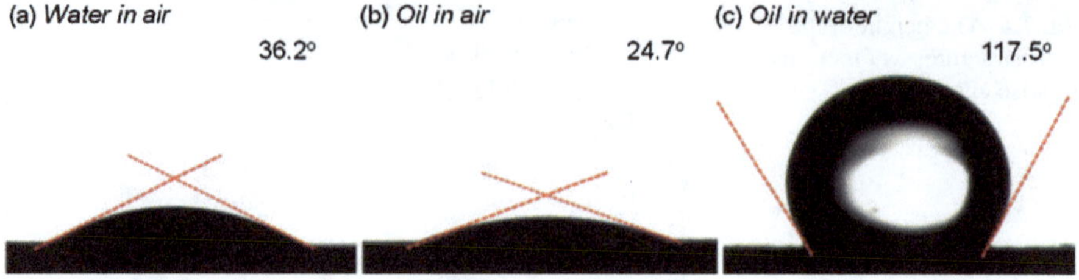

Fig. 7.5 **a** the water contact state in air on flat silicon surface. **b** The oil contact state in air on flat silocon surface. **c** The oil contact model underwater on flat silicon surface

hydrophilic, or more oleophilic or superoleophilic one by introducing the rough hierarchical micro/nanostructures. Figure 7.6a shows the contact angles (CA) as functions of the pulse energy for inducing microcones and micromolars with different roughness. It can be seen that the water CA decreased with the increasing pulse energy, from 24.5° to 3.2° for microcones, 12.3° to 2.6° for micromolars. Due to the hydrophilicity, the water can completely contact the rough surface, as shown in the schematic diagram of Fig. 7.6a, forming the Wenzel contact state.

According to Wenzel's model, the hydrophilic surfaces will be more hydrophilic with the increase of roughness. The CA measurement also shows that the structured silicon surfaces fabricated in sucrose solution are more hydrophilic than that in ethanol. Meanwhile, we find that scanning space also influence the water CA to a small degree, namely the water droplet has smaller CA at smaller scanning space. As seen from Fig. 7.6b, all the oil CAs on the structured silicon surface are less than 10° due to the lower surface tension, showing superoleophilicity. Similar to the case in Fig. 7.6a, the oil contact angle is decreased with the increasing pulse energy and the decreasing scanning space. It's worth mentioning that the water droplet can quickly spread on the micromolars induced at pulse energy above 0.18 mJ in sucrose solution, and finally the contact angle nearly reaches 0, showing extremely superhydrophilicity.

The properties of superhydrophilicity and superoleophilicity are crucial for preparing multi-functionalized surfaces. To this end, a fog collector constituted by three concentric rings spaced by unprocessed areas (Fig. 7.6c) is designed and placed into the artificial fog flow condition to reveal the enhanced fog harvesting. As time prolonged, the micro-droplets from fog are gathered to gradually form larger droplets, randomly distributed on the unprocessed locations. On the contrary, the fog droplet was prior to condense on the processed areas, and then fog water gradually coalesced and covered the concentric rings driven by the wettability difference during the fogging process. The continually collected fog water completely covered the three rings in the end, forming continuous waters. The volatilization behavior of the water droplets, which is an inverse process of the fog collection process, was also investigated (Fig. 7.6d). It is observed that the tiny water drops on the unprocessed areas is preferentially volatilized, while the water on the processed ones can keep the morphology primly. This investigation indicates that our fog collector can act as a

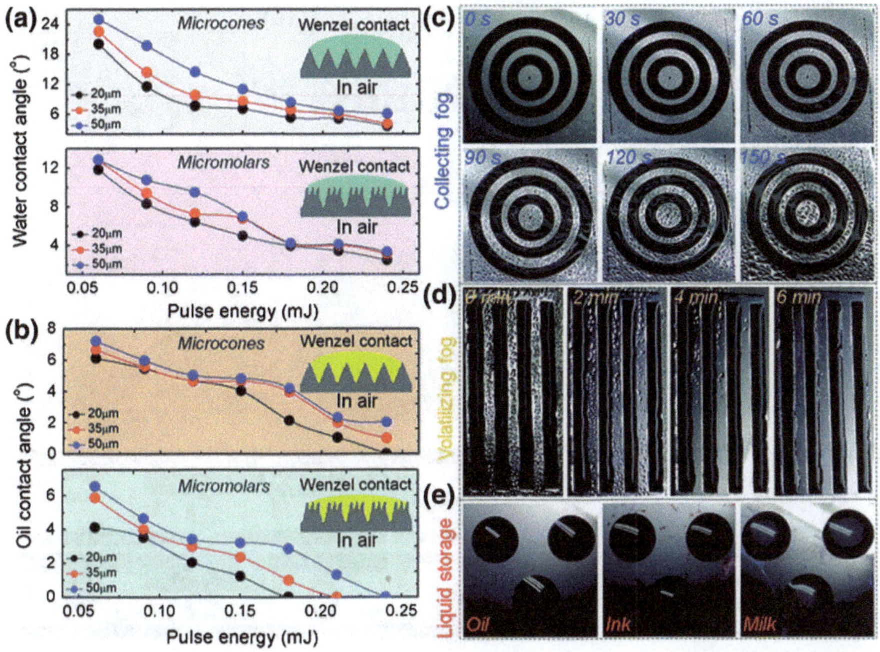

Fig. 7.6 The superhydrophilicity/superoleophilicity of the microcones and micromolars. **a** The water contact angle of microcones and micromolars. All contact angles decreased with the increasing pulse energy. **b** The oil contact angles of microcones and micromolars. The processed silicon surfaces are superoleophilic. The insert images in (a) and (b) were the illustration of the Wenzel contact model. **c** Sequential optical images of fog collection on microcones-constituted three concentric rings. **d** The volatilization process of the water collected by four 4 cm long lines consisting of micromolars arrays. Both the start time was set as 0 for (c) and (d). **e** The static storage for oil, red ink, and milk. Due to the superhydrophilicity and superoleophilicity, the liquid can be firmly fixed at the processed areas.

potential fog harvesting apparatus with an enhanced efficiency, which may find applications at the arid regions with freshwater scarcity. Due to the superhydrophilicity and superoleophilicity, the as-prepared silicon surfaces can be used for liquid storage. As shown in Fig. 7.6e, oil, red ink, and milk can be fixed at the processed three circular areas with 6 mm diameter, orderly. None of the droplets overflow from the designated circular regions, displaying ultrahigh adhesion.

7.4.2 Underwater Superoleophobicity and Oil Droplet Manipulation

When the hydrophilic silicon surface is immersed in water, water molecules can be trapped in the formed micro/nanostructures, developing a composite water/solid

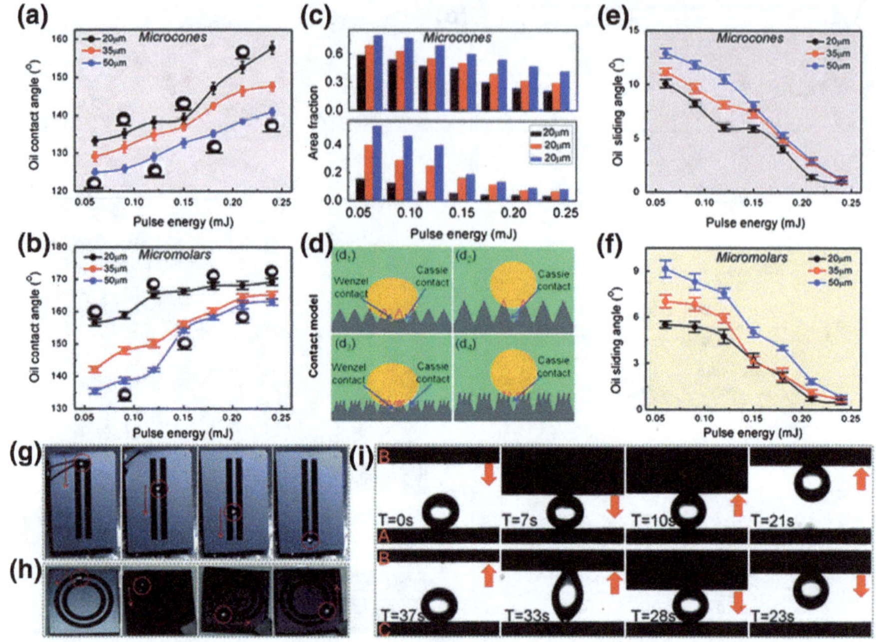

Fig. 7.7 The underwater superoleophobicity and the oil droplet manipulation. **a** and **b** show that the microcones and micromolars are underwater superoleophobic and the oil contact angles increased with the increasing pulse energy. **c** The calculated area fractions defined as the ratio of actual contact area by the oil droplet to the whole area of the microstructures. It is demonstrated that the area fractions decreased with the increasing pulse energy. **d** The detailed contact model for underwater oil. When the contact angle is smaller, the contact model is metastable Wenzel/Cassie hybrid states. However, if the contact angle is larger, the contact model is Cassie state. **e** and **f** are the oil sliding angles on microcones and micromolars as functions of pulse energy. **g** and **h** The directional transport of oil along the set path. **i** Process of transferring oil droplet from the A surface to C surface via B surface. Utilizing the differences of adhesive force, the structured surfaces can be used as "mechanical hand" to transfer the oil droplet from lower adhesive surface to higher one

interface. For the incompatibility, the trapped water acts as an insulation layer to prevent the oil droplet from permeating into the micro/nanostructures. In this case, the oil droplet can only contact the top of the induced micro/nano structures, hence the processed silicon surface forms an oleophobic and even superoleophobic interface. Figure 7.7a, b indicate the oil contact angles for microcones and micromolars as functions of pulse energy, respectively. It is easy to see that the oil contact angles are increased with the increasing pulse energy, and also they possesses a larger contact angle at smaller scanning space. By comparison, it is found that the micromolars endow the oil droplet with larger contact angle than the microcones. Furthermore, the state of the oil droplet can be described by the Cassie mode and the contact angle can be expressed as Eq. 6.2.

The calculated area fraction results are shown in Fig. 7.7c. Obviously, the value of f is decreased with the increasing pulse energy and decreasing scanning space.

In addition, it is observed from Fig. 7.7a, b that some measured contact angles are below 150°. In this case, the oil droplets are in metastable Wenzel/Cassie hybrid states, as shown in the schematic diagram of Fig. 7.7d_1, d_3. Conversely, when the contact angle is larger than 150°, the contact is the Cassie model, which represents a non-wet-contact mode. The oil contacts the upper end of the microstructures, leaving the lower end immersed in water (Fig. 7.7d_2, d_4). Especially, the minimum values for f are 0.206 and 0.081 for microcones and micromolars, endowing the structured surface with oil CA of 157.76° and 169.21°. The oil sliding behaviors are also studied (Fig. 7.7e, f). Similar to the case discussed above, the sliding angles are also decreased with the increasing pulse energy and decreased scanning space. The microcones and micromolars can modulate the oil sliding angle from 0.91° to 12.88°, and 0.52° to 9.14°, respectively. Generally, the small sliding angle signifies low adhesion [5].

The modification of the underwater wettability plays an important role in the oil manipulation. Figure 7.7g, h indicate the controlling of the moving behavior of the oil droplet. By utilizing the adhesion contrast between the treated and untreated area, the oil droplet can move along the setting path without loss via slightly tilting the sample (Fig. 7.8). In addition, the structured surfaces with controllable adhesion have important potential application in the transfer of oil droplets. Figure 7.7i shows an example of transferring a 5 μl oil droplet from the low adhesive A surface to the high adhesive C surface via using medium adhesive B surface as a "mechanical hand" [10].

7.4.3 The Anisotropic Wetting and Sliding Behavior

In the previous discussion, the scanning space (D) was set as 20, 35, and 50 μm, which can match up with the size of focused laser beam, therefore, there is hardly any untreated space between the scanning lines, making the induced microstructures overlap with each other. In this case, the wetting and sliding of the oil in water are isotropic. When increasing the scanning space, blank area appears and gradually grows wider as shown in Fig. 7.10a, b, which are prepared at pulse energy of 0.25 mJ. By contrasting the preparation in ethanol and sucrose solution, it is indicated that laser can scan the silicon surface to form the wider scanning lines with the assistance of sucrose solution than ethanol (Fig. 7.9) [5]. The increase of the scanning space results in the difference between the contact angles measured parallel or perpendicular to the scanning lines. Figure 7.10c, d reveal the shapes of a 5 μl oil droplet on the treated silicon surface with D of 80 and 200 μm, along the directions perpendicular and parallel the scanning lines, respectively. Seen from Fig. 7.10c for microcones with $D = 80$ μm, the values for $CA_{\perp}$ and $CA_{\parallel}$ are 153.92° and 151.83°, declaring a small degree of wetting anisotropy for the two directions. However, for $D = 200$ μm, the measured $CA_{\perp}$ and $CA_{\parallel}$ are 138.08° and 124.36°, hence the difference of the CA values for the two directions is 13.72°. Similarly, for micromolars, at $D = 80$ μm, the $CA_{\perp}$ and $CA_{\parallel}$ are 166.29° and 162.72°, and at $D = 200$ μm the values are 143.81° and 131.52°, respectively (Fig. 7.10d). We also systematically investigated the influence

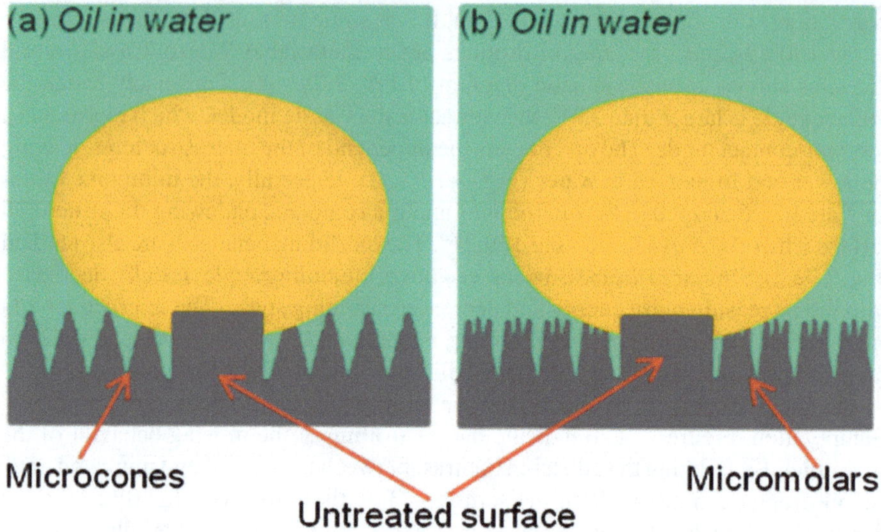

Fig. 7.8 Schematic diagram of underwater oil along the set path. The untreated silicon is oleophobic, while the treated areas **a** microcones and **b** micromolars are superoleophobic. By taking advantage of this property, we have designed a path composed of untreated fine line and treated thick lines on both sides. In this way, the fine line can grasp the oil droplet and control it to move along the setting path, and the ultralow adhesion force of the processed lines on both sides ensure the oil droplet to move without loss

of *D* on the values for CAs of oil in the perpendicular and parallel directions on both microcones and micromolars (Fig. 7.10e). It can be seen that although both the $CA_{\perp}$ and $CA_{\parallel}$ are decreased with the increasing *D*, $CA_{\perp}$ is consistently larger than $CA_{\parallel}$. Furthermore, $CA_{\parallel}$ is initially approximate to $CA_{\perp}$, however, with *D* increasing, the $CA_{\parallel}$ is decreased much faster for the more and more obvious anisotropic structure.

To investigate the influence of *D* on the anisotropic sliding of the oil droplets, the SAs perpendicular and parallel to the scanning lines (Fig. 7.10f) were measured. Figure 7.10g, h indicate the snapshots of an oil droplet rolling on tilted microcones and micromolars silicon surface with *D* of 80 and 200 μm, respectively. At *D* of 80 μm, the $SA_{\perp}$ and $SA_{\parallel}$ are 1.13° and 2.87° for microcones, 0.81° and 0.84° for micromolars, showing small sliding anisotropy. However at *D* of 200 μm, the measured $SA_{\perp}$ and $SA_{\parallel}$ are 48.24° and 58.79° for microcones, inducing sliding anisotropy is 10.55°. While for micromolars, the $SA_{\perp}$ is 21.45° and $SA_{\parallel}$ is 44.29°, which indicates that the sliding anisotropy reaches 22.84°. The relationship between the anisotropic sliding and *D* for the microcones and micromolar was systematically studied (Fig. 7.10i). From Fig. 7.10i, it is observed that the SAs in the parallel direction were always smaller than those in the perpendicular direction. This anisotropic sliding is caused by the different three-phase contact line (TCL) [11]. By comparison, the discontinuous and long TCL perpendicular to the scanning lines can give

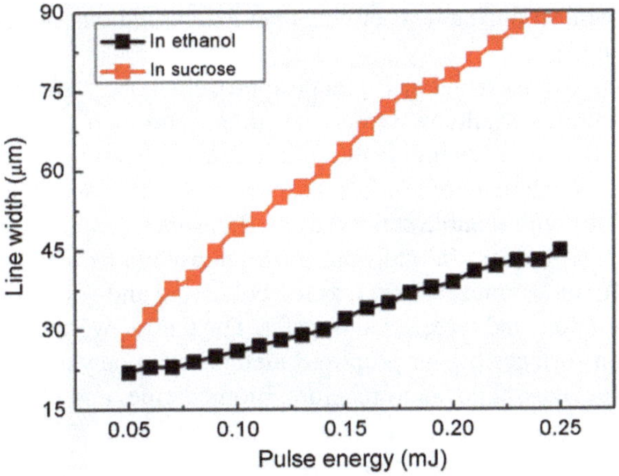

Fig. 7.9 The line width of the scanning lines as functions of the pulse energy

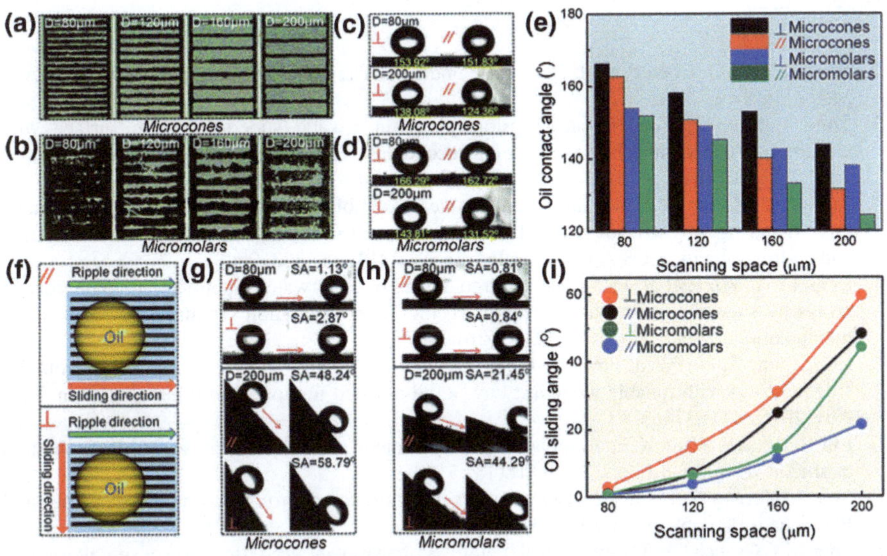

Fig. 7.10 The anisotropic wetting and sliding behavior. **a** and **b** are the optical microscope images of the silicon surfaces prepared by ethanol and sucrose solution assisted fs laser irradiation under pulse energy of 0.25 mJ and scanning space of 80, 120, 160, and 200 μm, respectively. **c** and **d** The optical images of the anisotropic contact state of 5 μl oil which were measured perpendicular and parallel to the scanning lines composed by microcones and micromolars, respectively. **e** The statistical results of the oil anisotropic contact angles with the increasing scanning spaces. **f** The schematic illustration of oil droplet slides perpendicular and parallel to the scanning lines. **g** and **h** The snapshots of oil droplet rolling on the microcones and micromolars surfaces with scanning spaces of 80 and 200 μm, along the parallel and perpendicular directions, respectively. **i** The oil anisotropic sliding angles as a function of the scanning space

rise to more pronounced hysteresis effects than the continuous short ones along the parallel direction [5].

In this section, we have developed a facile approach of one-step assembling microcones and micromolars on silicon surfaces by ethanol and sucrose solution-assisted fs laser irradiation. By simply changing the solutions and adjusting the pulse energy, the height of the microcones and micromolars can be arbitrarily controlled, which endow the silicon surfaces with tunable superhydrophilic, superoleophilic, and underwater superoleophobic properties. In addition, these micro/nanostructured surfaces also show many useful multifunctions, such as fog collecting and volatilizing, liquid storage and transportation, and directional transfer. The unique wetting properties of the structured silicon surfaces by our proposed method open new applications in fields of microfluidic devices, fluid microreactors, biomedicine, and chemical/biological sensors [5].

References

1. Yong J, Yang Q, Chen F, et al. Reversible underwater lossless oil droplet transportation. Adv Mater Interfaces. 2015;2(2).
2. Yong J, Chen F, Yang Q, et al. Bioinspired underwater superoleophobic surface with ultralow oil-adhesion achieved by femtosecond laser microfabrication. J Mater Chem A. 2014;2(23):8790–5.
3. Li G, Li J, Zhang C, et al. Large-area one-step assembly of three-dimensional porous metal micro/nanocages by ethanol-assisted femtosecond laser irradiation for enhanced antireflection and hydrophobicity. ACS Appl Mater Interfaces. 2014;7(1):383–90.
4. Li G, Lu Y, Wu P, et al. Fish scale inspired design of underwater superoleophobic microcone arrays by sucrose solution assisted femtosecond laser irradiation for multifunctional liquid manipulation. J Mater Chem A. 2015;3(36):18675–83.
5. Li G, Zhang Z, Wu P, et al. One-step facile fabrication of controllable microcone and micromolar silicon arrays with tunable wettability by liquid-assisted femtosecond laser irradiation. RSC Adv. 2016;6(44):37463–71.
6. Liu H, Chen F, Wang X, et al. Influence of liquid environments on femtosecond laser ablation of silicon. Thin Solid Films. 2010;518(18):5188–94.
7. Yan Z, Chrisey DB. Pulsed laser ablation in liquid for micro-/nanostructure generation. J Photochem Photobiol C. 2012;13(3):204–23.
8. Shen MY, Crouch CH, Carey JE, et al. Femtosecond laser-induced formation of submicrometer spikes on silicon in water. Appl Phys Lett. 2004;85(23):5694–6.
9. Baldacchini T, Carey JE, Zhou M, et al. Superhydrophobic surfaces prepared by microstructuring of silicon using a femtosecond laser. Langmuir. 2006;22(11):4917–9.
10. Bi H, Xie X, Yin K, et al. Spongy graphene as a highly efficient and recyclable sorbent for oils and organic solvents. Adv Func Mater. 2012;22(21):4421–5.
11. Yong J, Yang Q, Chen F, et al. A simple way to achieve superhydrophobicity, controllable water adhesion, anisotropic sliding, and anisotropic wetting based on femtosecond-laser-induced line-patterned surfaces. J Mater Chem A. 2014;2(15):5499–507.

Chapter 8
Micropore-Arrayed Ultrathin Aluminum Foil for Oil/Water Separation and Particle Filtration

The separation of oil/water mixtures has become a global concern due to the serious oil pollution caused by discharging industrial oily wastewater, offshore oil accidents and marine transportation [1–4]. Because oil/water separation is an interfacial issue which is controlled by the chemical composition and geometrical structures, designing and constructing functional interfacial materials possessing different wettability to oil and water is proven to be an effective strategy for wastewater treatment [4–9]. Among these functional materials, porous sponges and foams [1, 3, 6, 9–14] have been extensively used for oil/water separation. Although sponges and foams are characterized by light weight, great porosity, good elasticity, and high-efficiency in oil/water separation, these materials are limited to the design and control the internal structures which are mainly determined by the parent skeletons [15]. In addition, these porous materials have limited adsorption capacity for absorption saturation makes them can't absorb more oil. Moreover, their fabrication usually needs complex chemical processing, which bring in serious pollution of the environment. Besides the adsorbent sponges and foams, many meshes have been constructed as alternative strategy and have been successfully applied in oil/water separation [16–21]. For example, Chen et al. [22] first prepared simultaneous superhydrophobic and superoleophilic Teflon (PTFE) coated mesh film for separation of oil and water. Zhang et al. [19] reported a nanowire-hair microstructures covered $Cu(OH)_2$ copper mesh which had a high separation capacity for both water-rich immiscible mixtures and dispersed oil-water mixtures. However, their preparation is also performed in chemical conditions which may cause secondary pollution to the environment. Gao et al. [21] developed a dual-scaled porous nitrocellulose membrane which had high oil/water separation efficiency by a facile perforating method. Although this process is scalable, fast, low-cost, and high separation efficiency, the formed pores are nonuniform and uncontrollable, which hinder their extensive use. Additionally, most of the literatures do not concern the solution to the continuous separation for the gravity-driven removal of heavy or light oils (compared to the density of water). Furthermore, these meshes are restricted to the oil/water separation, not refer to other functions. Hence, it is of great significance to seek novel functional materials and prepare regular, controllable structures with multifunctions in a simple, environment-friendly, and scalable approach [23].

G. Li, *Bionic Functional Structures by Femtosecond Laser Micro/nanofabrication Technologies*, Springer Theses, https://doi.org/10.1007/978-981-13-0359-3_8

Here, for the first time, we report the one-step femtosecond laser irradiation to induce large-area regular micropore arrays on ultrathin aluminum foils, which have thin thickness (~25 μm), light weight, low cost, good plasticity, and have been commonly applied to packaging, cooking, geochemical sampling, art and decoration and so on. The micropore-arrayed aluminum foils show underwater superoleophobic ability which is responsible for oil/water separation function. By skillfully designing the shape and structure of the aluminum foil oil/water separator, the oil and water can be separated efficiently and continuously. In addition, the uniform micropores were used to filtrate the particles with different diameters, showing multifunctions. This work provides a new pathway for controllable construction of aluminum foil-based micropore arrays with high separation efficiency and extends their applications to broader fields besides water treatment [23].

8.1 The Preparation of the Micropore-Arrayed Aluminum Foils

Ultrathin aluminum foil with regular micropore arrays were fabricated by a femtosecond laser perforating method, and the schematic fabrication process is shown in Fig. 8.1. By setting the laser pulse energy of 50 μJ and perforating space of 20 μm, an area of $40 \times 40\ \text{mm}^2$ were obtained (Fig. 8.2). From the optical camera photo taken under the inclined incidence of white light, it is seen that the micropore-arrayed aluminum foils presents brilliant iridescence due to the optical diffraction effect. In addition, bright iridescence can also be formed by the irradiation of white LED from the behind (Fig. 8.3) [23]. In Fig. 8.3a, the four iridescent patterns are taken by distances of 0, 5, 10, 15 cm, from the foils to the camera, respectively. While in Fig. 8.3b, the iridescent patterns are taken by different angles to the optical axis, and the included angles are 0°, 5°, 10°, and 15°, respectively.

From the low-magnification SEM image and the transmission microscope photographs (Fig. 8.4) [23], it is clearly seen that uniform micropore arrays are neatly arranged on aluminum foil surface. The magnified SEM image exhibited that the pore size is about 2.4 μm, and the rim is covered with rough nanostructures. The uniformity of the micropores is studied by systematically analyzing the pore diameters and the eccentricity, and the results are shown in Fig. 8.5 [23].

The mean diameter of the pores (Fig. 8.5b) and the eccentricity of the each pore (Fig. 8.5c) can be obtained by averaging the X and Y directions in the blue area in Fig. 8.5a. It can be seen that all the eccentricity is less than 0.1, which means that regular micropores are arrayed on the aluminum foil surface.

Figure 8.6a expresses the representative SEM images for micropore intervals of 40, 60, and 80 μm, which can be adjusted by altering the perforating space of the laser beam. Additionally, three pores with diameters of 8.2, 19.1, and 32.2 μm are demonstrated in Fig. 8.6b, and the size distribution is analyzed in Figs. 8.6c and 8.7. The results indicate that the pores are uniformly distributed. In addition, the

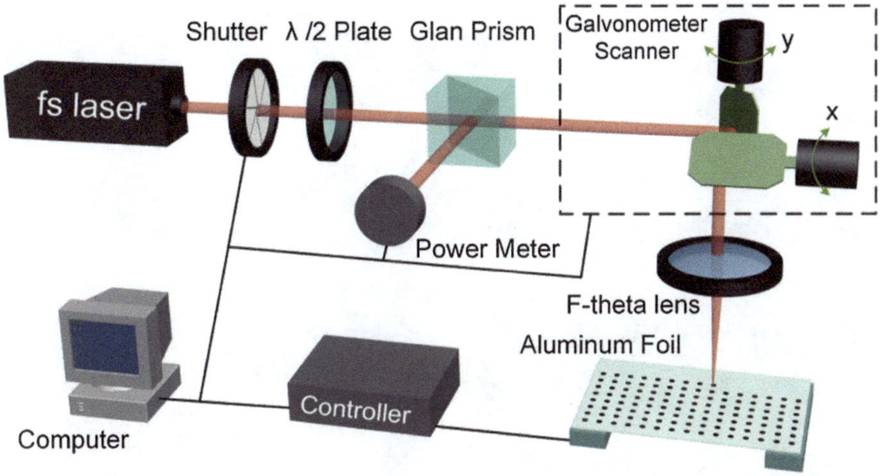

Fig. 8.1 The schematic diagram of fabricating regular micropore-arrayed aluminum foil

Fig. 8.2 Optical images of the as-prepared aluminum foil

preparation of a series of micropores with diameters ranged from 2.4 to 32 μm is also studied by changing the laser pulse energy and pulse numbers, and the results are shown in Fig. 8.6d.

From Figs. 8.2 and 8.7, it is clearly indicated that regular micropore arrays with diverse diameters and intervals can be easily prepared on aluminum foil surfaces just by simply adjusting the laser processing parameters without other complex process. In addition, this process is fast. For example, a 10×10 mm^2 membrane with micropore size of ~15 μm and interval of 60 μm can be processed in less than 6 min.

Fig. 8.3 Bright iridescence formed by the regular micropore arrays due to the optical diffraction effect. **a** The four iridescent patterns taken by different distance from the foils to the camera. The distances are 0, 5, 10, 15 cm, respectively. **b** The iridescent patterns taken by different angles to the optical axis, and the included angles are 0°, 5°, 10°, and 15°, respectively. All the samples are irradiated by white LED from the behind

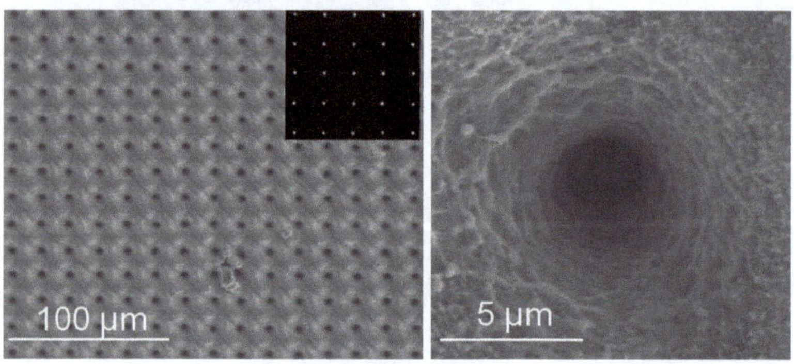

Fig. 8.4 SEM images of the as-prepared aluminum foil

8.2 The Wettability of the Micropore-Arrayed Aluminum Foils

The unprocessed aluminum foil is hydrophilic and the water contact angle is 53.9°. The microsized pores in combination with nanoscale roughness endow a hierarchical composite structure to the aluminum foil that possesses superwetting behaviors. For example, the aluminum foil with pore diameter of 2.4 μm and interval of 20 μm obtains a water contact angle of 7.8° (Fig. 8.8a) [23]. When the

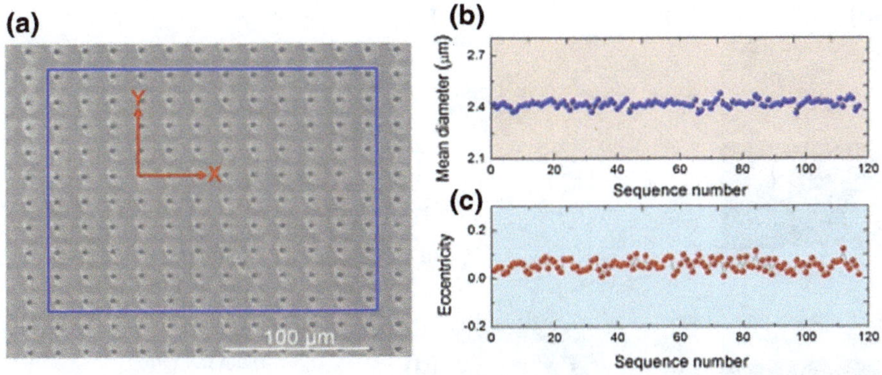

Fig. 8.5 Uniformity analysis of the micropores formed by femtosecond laser perforating. **a** is the SEM image of the micropore arrays. The interval of the adjacent pores is 20 μm. **b** is the mean diameter of the pores by averaging the X and Y directions in the blue area in (a). **c** is the analysis of the eccentricity of the each pore in the blue area in (a). It can be seen that all the eccentricity is less than 0.1, which means that the pores are very equivalent to roundness. From (b) and (c), it is indicated that regular micropores are arrayed on the aluminum foil surface

as-prepared aluminum foil is immersed in water, its wetting property is reverted to be superoleophobic. Typically, for C_8H_{18} whose density is less than water, and for $C_2H_4Cl_2$ which is heavier than water, the contact angles are 153.5° (Fig. 8.8b) and 153.1° (Fig. 8.8c), respectively.

The hydrophilicity and underwater oleophobicity of the aluminum foil prepared under different laser processing parameters are systemically studied and the results are summarized in Fig. 8.9 [23]. Figure 8.9a is the water contact angle on micropores with different diameters and intervals. It is indicated that most of the contact angles are less than 10°, showing superhydrophilic nature. In addition, the contact angles are decreased with the increasing micropore diameters and decreasing intervals. Figure 8.9b, c are the C_8H_{18} and $C_2H_4Cl_2$ contact angles on micropore arrays. It can be seen that all the contact angles are larger than 150°, exhibiting underwater superoleophobic property. From Fig. 8.9b, c, it is observed that the oil contact angles are increased with the increasing micropore diameters and decreasing intervals. This study illustrates that the wettability can be accurately controlled by changing the micropore diameters and intervals.

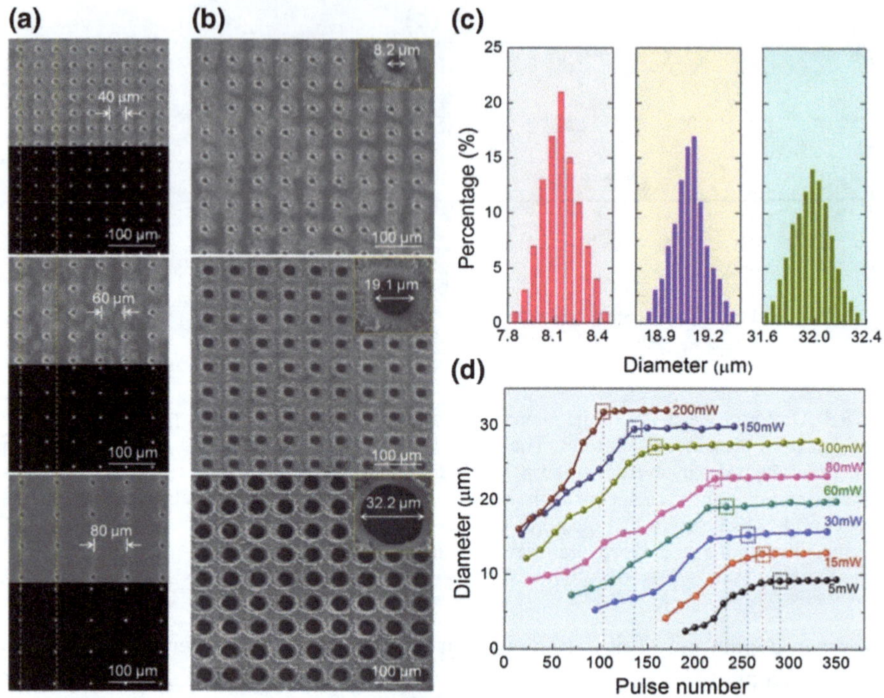

Fig. 8.6 The precise control of regular micropore arrayes with diverse intervals and diameters. **a** The SEM and transmission optical microscope images of the micropores with intervals of 40, 60, and 80 μm, from top to bottom, respectively. **b** The representative SEM images of the micropores with diameters of 8.2, 19.1, and 32.2 μm, respectively. **c** Distribution statistics of the three pore diameters in (b), showing good uniformity. **d** The diameters as a function of laser pulse energy and pulse number

8.3 The Oil/Water Separation by the Micropore-Arrayed Aluminum Foils

The opposite wetting properties for water and oil are supposed to be applied for controllable oil/water separation. Firstly, the oil/water mixtures were prepared and the results are shown in Fig. 8.10. From Fig. 8.10a–c, it can be seen that the undyed water, C_8H_{18}, and $C_2H_4Cl_2$ are colorless transparent. The water presents blue color after it is dyed with methylene blue. Meanwhile, the C_8H_{18} and $C_2H_4Cl_2$ show red color after dyed with Sudan IV. It is observed from Fig. 8.10d–f that the layered oil and water is formed due to C_8H_{18} is lighter than water while $C_2H_4Cl_2$ is heavier than water [23].

Several proof-of-concept studies were performed to test the oil/water separation capacities of the as-prepared aluminum foil. Here we choose the C_8H_{18} and $C_2H_4Cl_2$ as the target oils, and the photographs of mixtures are shown in Fig. 8.11a. The C_8H_{18}/water separation experiment is operated, and the schematic diagram is

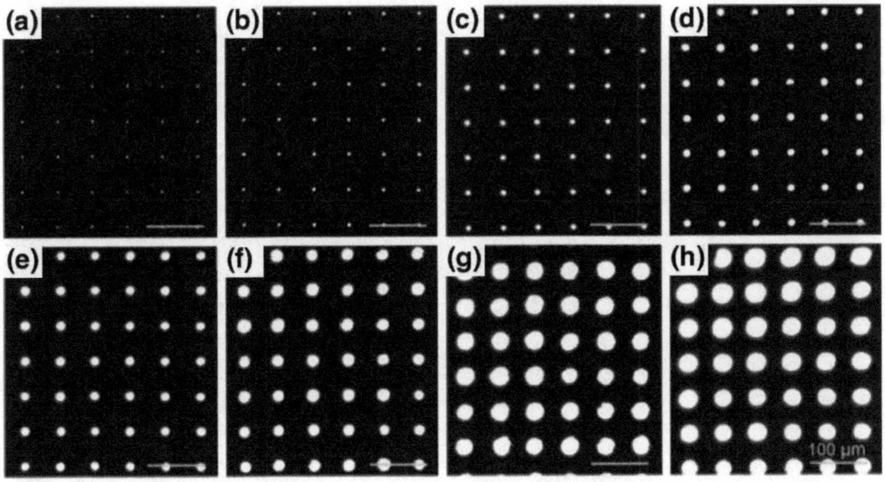

Fig. 8.7 Optical microscope images of the micropores. From **a** to **h**, the diameters are 2.4, 4.1, 8.2, 12.2, 15.9, 19.1, 27.7, and 32.2 μm, respectively. The results indicate that not only the pores are uniformly distributed, but also the micropores can be tuned by changing the laser pulse energy and pulse numbers

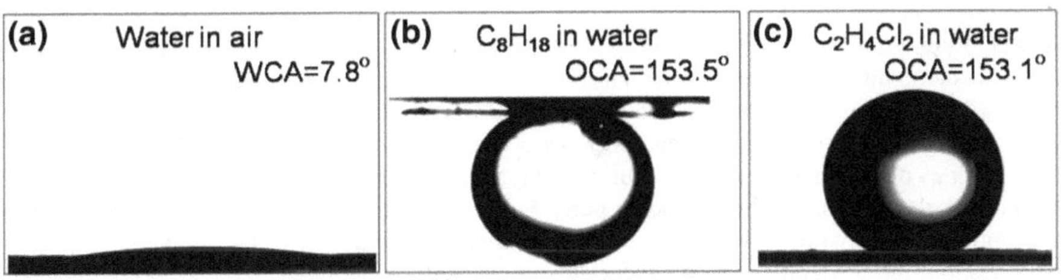

Fig. 8.8 The contact-angle images of the aluminum foil, demonstrating superhydrophilic (water contact angle is about 7.8° (a)), underwater superoleophobic (contact angle of C_8H_{18} and $C_2H_4Cl_2$ are 153.5° (b), and 153.1° (c), respectively)

displayed in Fig. 8.11b. As shown in Fig. 8.11c, a piece of aluminum foil with pore diameter of 8.2 μm and interval of 60 μm was horizontally fixed between a glass tube and conical flask to act as a separating mesh, which is a common installation method reported in previous literatures. A beaker of 20 ml C_8H_{18}/water mixture was poured onto the aluminum foil, and then the separation was proceeded by the force of gravity. Due to the superhydrophilicity, the water (blue color) rapidly permeates the micropores and then in pours the conical flask, whereas the C_8H_{18} (red color) is blocked in the upper tube for the superoleophobicity of the aluminum foil. The C_8H_{18}/water separation is completed in less than 13 s and no external force other than gravity is needed during the separation process. In this way, 1.54×10^5 L C_8H_{18}/water mixture can be separated by using an aluminum foil with processing area of 1 m^2 in one hour. This kind of separator can be used to effectively separate a

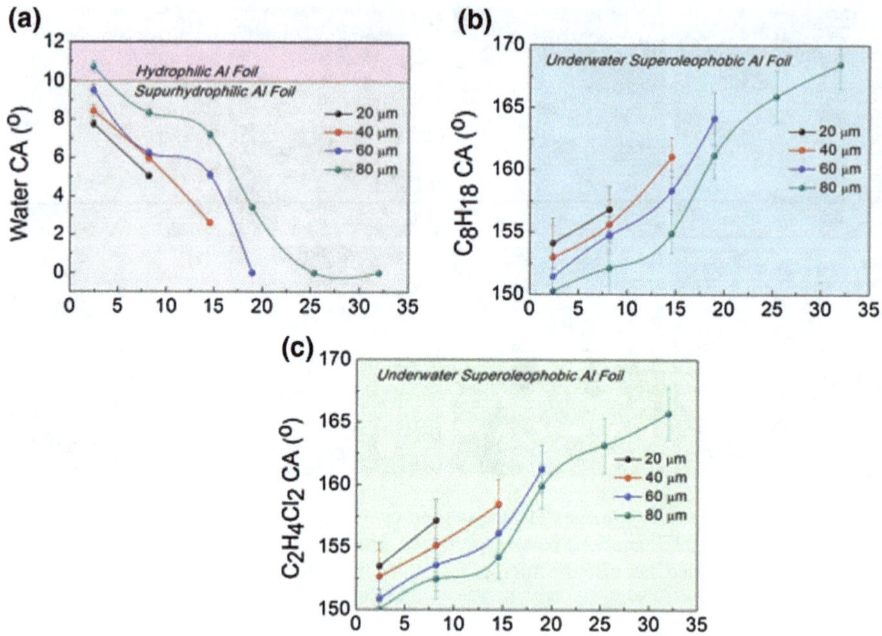

Fig. 8.9 Systematic study of the wettability of the micropore arrays. **a** is the water contact angle on micropores with different diameters and intervals. It is indicated that most of the contact angles are less than 10°, showing superhydrophilic nature. In addition, the contact angles are decreased with the increasing micropore diameters and decreasing intervals. **b** and **c** are the C_8H_{18} and $C_2H_4Cl_2$ contact angles on micropore arrays. It can be seen that all the contact angles are larger than 150°, exhibiting underwater superoleophobic property. From (b) and (c), it is observed that the oil contact angles are increased with the increasing micropore diameters and decreasing intervals. This study illustrates that the wettability can be accurately controlled by changing the micropore diameters and intervals

mixture of light oil (whose density is less than water) and water. However, this device is not suitable for the separation heavy oil (heavier than water) and water mixtures for the heavy oil settling below the water, forming an insulation layer between the water and the aluminum foil which prevents the water from maintaining contact with the aluminum foil. Although some works have mentioned the separation heavy oil and water mixtures with this method, it is speculated that the assistant external forces, such as stirring and shaking are needed to churn the mixtures to let the water contact with the aluminum foil [23].

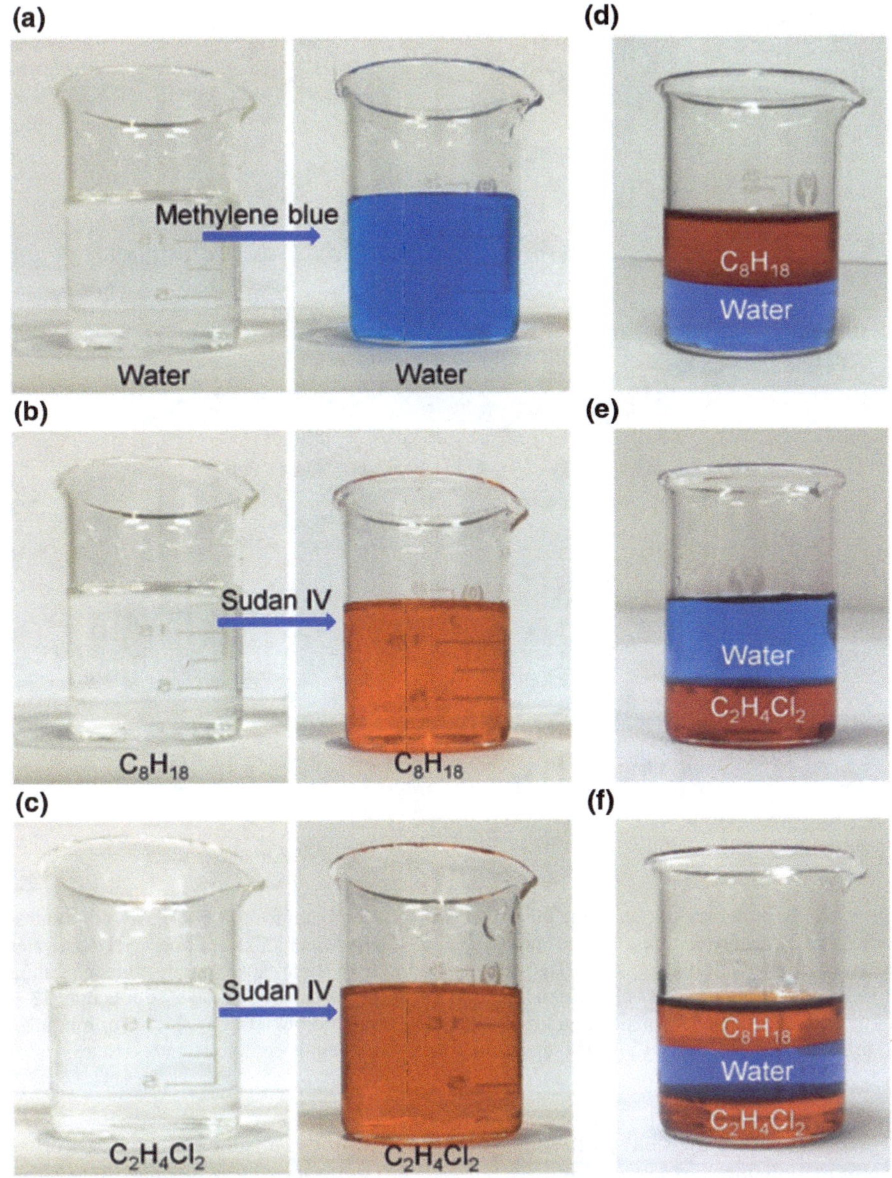

Fig. 8.10 Preparation of the oil/water mixtures. From (a) to (c), it can be seen that the undyed water, C_8H_{18}, and $C_2H_4Cl_2$ are colorless transparent. The water presents blue color after it is dyed with methylene blue. Meanwhile, the C_8H_{18} and $C_2H_4Cl_2$ show red color after dyed with Sudan IV. It is observed from (d) to (f) that the layered oil and water is formed due to C_8H_{18} is lighter than water while $C_2H_4Cl_2$ is heavier than water

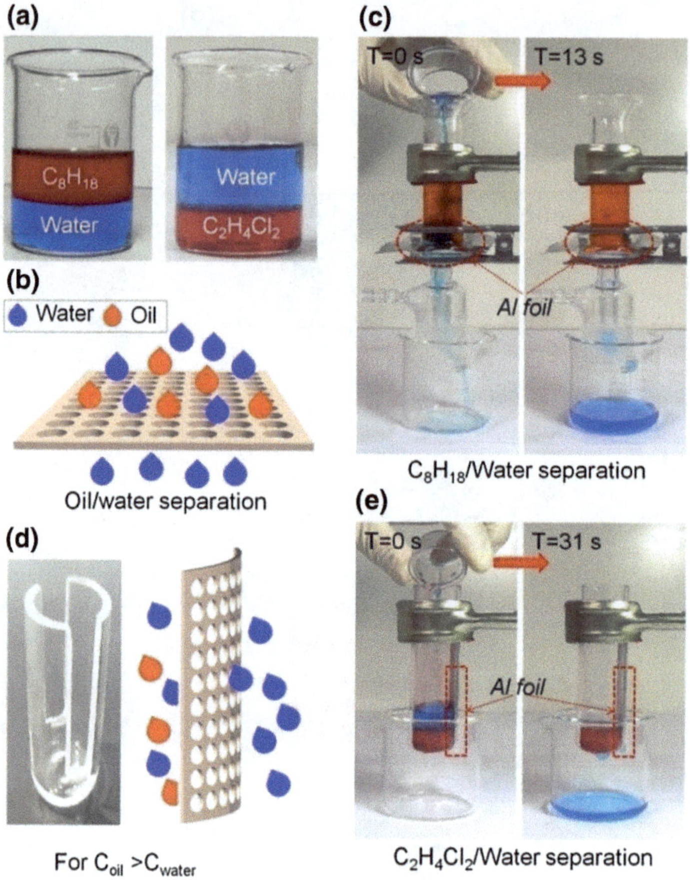

Fig. 8.11 Oil/water separation of the micropore arrayed aluminum foil for C_8H_{18}/water mixtures and $C_2H_4Cl_2$/water mixtures. **a** The photographs of C_8H_{18}/water mixtures and $C_2H_4Cl_2$/water mixtures. **b** Schematic diagram of C_8H_{18}/water separation. **c** The time-lapse images of the separation of C_8H_{18}/water mixtures. It is indicated that a 20 mL mixtures can be separated in less than 13 s. **d** The equipment and schematic diagram of separating heavy oil of $C_2H_4Cl_2$ and water mixtures. **e** Time-lapse images of C_8H_{18}/water separation, and this process is completed in less than 31 s

For separating the heavy oil and water mixtures in the absence of other external forces, a tube with a side opening which is covered with micropore arrayed aluminum foil is designed (Fig. 8.11d). When the mixture of heavy oil and water is poured into the upright tube, the water in the oil/water mixture can pass through the side opening although the oil settles below the water, thereby achieving oil/water separation only with gravity effect. The results show that a 20 ml heavy oil/water mixture can be separated in less than 31 s, which means that the accomplishment of separation of 6.5×10^4 L oil/water mixture by per m^2h (Fig. 8.11e) [23].

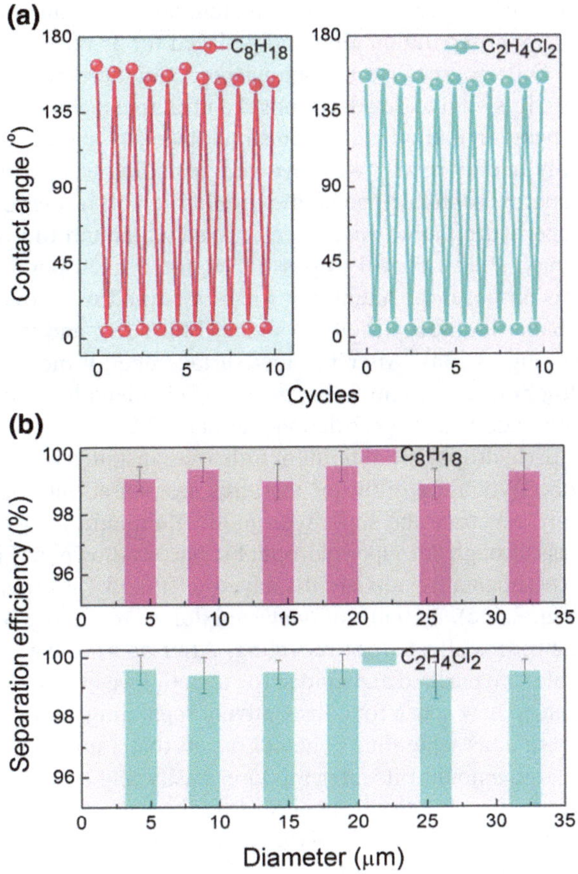

Fig. 8.12 **a** Cycling performance of micropore-arrayed aluminum foil for repeated use after ultrasonic cleaning. **b** The calculated separation efficiency of the aluminum foil with pore diameters ranging from 2.4 to 32 mm for both C_8H_{18}/water mixtures and $C_2H_4Cl_2$/water mixtures

Furthermore, this separator can separate not only heavy oil/water mixtures but also light oil/water mixtures. Besides, the aluminum foil can maintain its superhydrophilicity and underwater superoleophobicity after 10 separations and is easily cleaned using water or ethanol for reuse (Fig. 8.12a). The separation efficiency of the micropores with pore diameters from 2.4 to 32 μm was quantitatively evaluated by calculating the ratio of the oil before and after separation according to the following equation [18–23].

$$R(\%) = (1 - \frac{C_p}{C_o}) \times 100 \tag{8.1}$$

here, C_o and C_p refer to the oil concentration in the original oil/water mixture and the collected water, respectively. It is indicated that the separation efficiency of this series of micropores was calculated up to 99% for both C_8H_{18} and $C_2H_4Cl_2$ (Fig. 8.12b).

In addition, we found that the aluminum foil can still be used for high-efficiency oil/water separation after being placed for 35 days.

Oil resources are becoming more and more precious with their rapidly running out. It is vital to develop a novel and simple technique that can achieve the oil/water separation and oil collection simultaneously and continuously. To obtain the effect, only one more step was involved: installing a superhydrophilic aluminum foil and superhydrophobic one on the openings on both sides of the tube. This is because the superhydrophobic micropore-arrayed aluminum foil is underwater superoleophilic, which allows the oil to pass through the micropores and block off the water. For this purpose, the aluminum foil is modified with low surface energy *1H*, *1H*, *2H*, *2H*-perfluorodecyltriethoxysilane (PFDTES), and the schematic diagram is shown in (Fig. 8.13a). After the low surface energy modification, the original superhydrophilic aluminum foil (CA = 5.1°) is tuned to superhydrophobic (CA = 154.4°) and underwater superoleophobic one (CA = 9.5°). Then the superhydrophilic and superhydrophobic aluminum foils were installed on the openings on each side of the tube. When the oil/water mixtures are poured into the upright tube, the water can only penetrate the superhydrophilic aluminum foils micropores while the oil only pass through the superhydrophobic one, as shown in (Fig. 8.13b). The corresponding experimental results are displayed in (Fig. 8.13c). Just as expected, it is seen from (Fig. 8.13c), a 60 mL oil/water mixtures are rapidly separated in less than 28 s and the oil is collected for recycling. After pouring the oil/water mixtures, the oil and water are purified and guided to collecting vessel. In the meantime, the separator can free up new space for consecutively separating subsequent oil/water mixtures. Such continuous separating behavior means that a small separator can separate and purify a large amount oil/water mixtures easily and quickly. This novel strategy not only overcomes the disadvantages of traditional methods which can only separate light or heavy oil and water mixtures in small amounts, but also provides a possibility of recycling water and oil.

In addition, due to the fast and complete discharge of the water and oil from the separator (Fig. 8.13c), the oil/water separation can withstand the intrusion pressure of oils flowing through the Al foil, which is determined by the oil layer height and can be expressed experimentally and theoretically by Eq. (8.2) [18–21] and Eq. (8.3) [18–21]

$$P_{\text{exp}} = \rho g h_{\text{max}} \tag{8.2}$$

$$P_{theor} = 2\gamma_{OW} \cos\theta_{OW}/d \tag{8.3}$$

in Eq. (8.2), ρ, g, and h_{max} are the density of the oil, the acceleration of gravity, and the maximum height of oil layer that the aluminum foil can support. In Eq. (8.3), d refers to the diameter of the micropores (Fig. 8.14) [23]. Hence, this novel separator may open a new channel for the fast, continuous, and high-efficiency oil/water separation.

We also tested the maximum height of the water that the membrane can sustain. A membrane with pore size of ~13.1 μm and interval of 60 μm was fixed at the lower end of the glass tube. The tested oil is $C_2H_4Cl_2$ and the pressure is exerted by adding

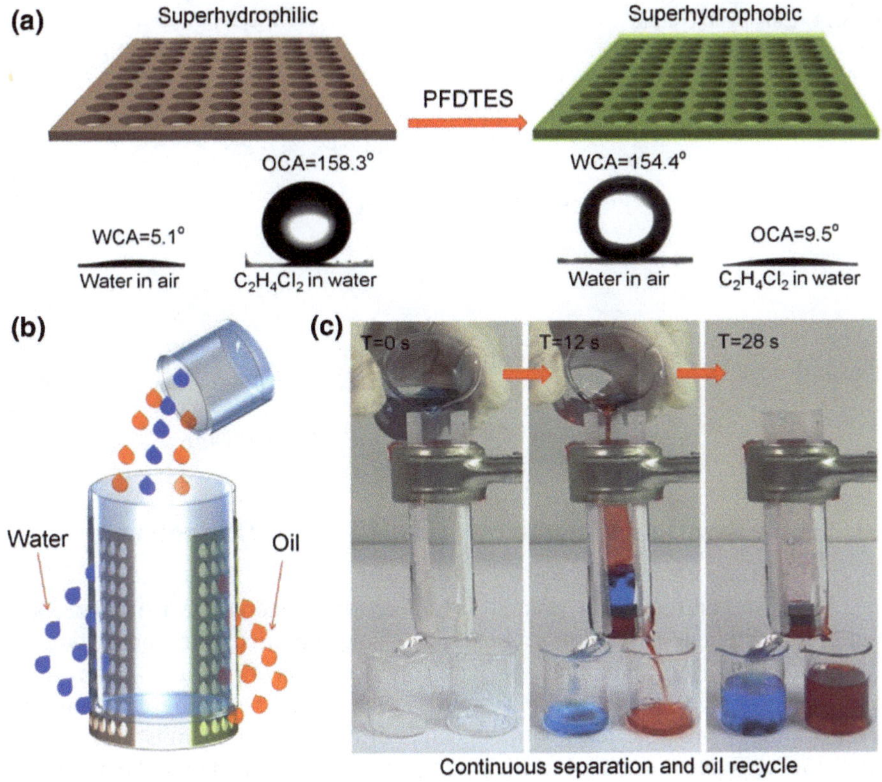

Fig. 8.13 The demonstration of the continuous separation of oil/water mixtures. **a** Schematic illustrations of fabrication of superhydrophobic aluminum foil by modifying the superhydrophilic one with low surface energy 1H, 1H, 2H, 2H-perfluorodecyltriethoxysilane (PFDTES). The modified aluminum foil is superhydrophobic (water contact angle of 154.42°) and underwater superoleophilic ($C_2H_4Cl_2$ contact angle of 9.47°). **b** The home-made devices for continuously separating oil and water mixtures. **c** Time-lapse images of continuous oil/water mixtures separation and the recycle for oil. A 60 mL oil/water mixtures are rapidly separated with foil area of 0.3×3 cm^2 in less than 28 s and the oil is collected for recycling. In this way, 23.8 L oil/water mixtures can be separated with 1 m^2 aluminum foil in less than 1 s

water into the upright glass tube. A membrane with pore size of ~13.1 μm and interval of 60 μm was fixed at the lower end of the glass tube. Before the test, the membrane is pre-wetted by spraying water. Due to the underwater superoleophobicity, the oil can't pass through the membrane. When the height of the water reaches ~110.6 cm, the oil can penetrate the superoleophobic membrane and flows into the beaker. This test is also conducted on the membrane with pore size of ~24.5 μm, and the obtained maximum height of water is ~21 cm. As for the membrane with the size of ~2.4 μm, the maximum water height can't be measured for the maximum length of the glass tube is 150 cm. Even so, we believe the maximum water height is far beyond 150 cm.

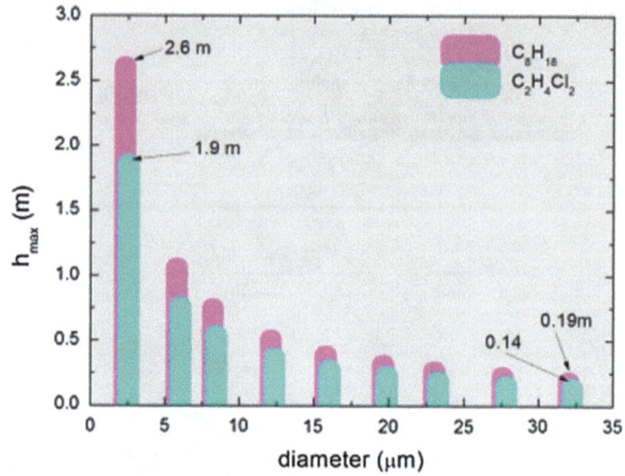

Fig. 8.14 Calculation of the maximum height of the oil layer which can be supported by the aluminum foil

8.4 The Filtration of Particles with Different Diameters

The regular and uniformly distributed micropore arrays endow the as-prepared aluminum foil with a new function, namely the enhanced particle filtration. In order to illustrate this function, aluminum foils with pore diameters of 13.2 and 7.4 μm, interval of 60 μm (Fig. 8.15a) are applied to filtrate the particles with different diameters. The filtration of SiO_2 particles with diameters of 20, 10, and 5 μm is illustrated in Fig. 8.15b, c. When the aqueous solution of particles is dropped onto the micropore arrayed aluminum foil, the particles whose sizes is less than the pore diameters can pass through the micropores under current. However, the particles which are bigger than the micropores will be retained above the aluminum foil. From experimental results exhibited in Fig. 8.15d, e, it can be seen that the sorted particles sizes are all 10 and 5 μm, respectively, indicating high sorting efficiency. We believe that the simplicity of this approach makes it quite appealing for not only particle filtration with various sizes, but also blood and cell separation, and even more lab-on-a-chip devices.

It's worth mentioning that due to the controllability and flexibility of the femtosecond laser micro/nanofabrication, diverse patterns, such as heart, TaiJi, and star-shaped patterns composed by micropore arrays can be simply and rapidly fabricated (Fig. 8.16a). Moreover, the good plasticity of the aluminum foil make it folded into diverse three-dimensional geometric patterns (Fig. 8.16b). All these advantages are not available to other traditional preparation methods and materials.

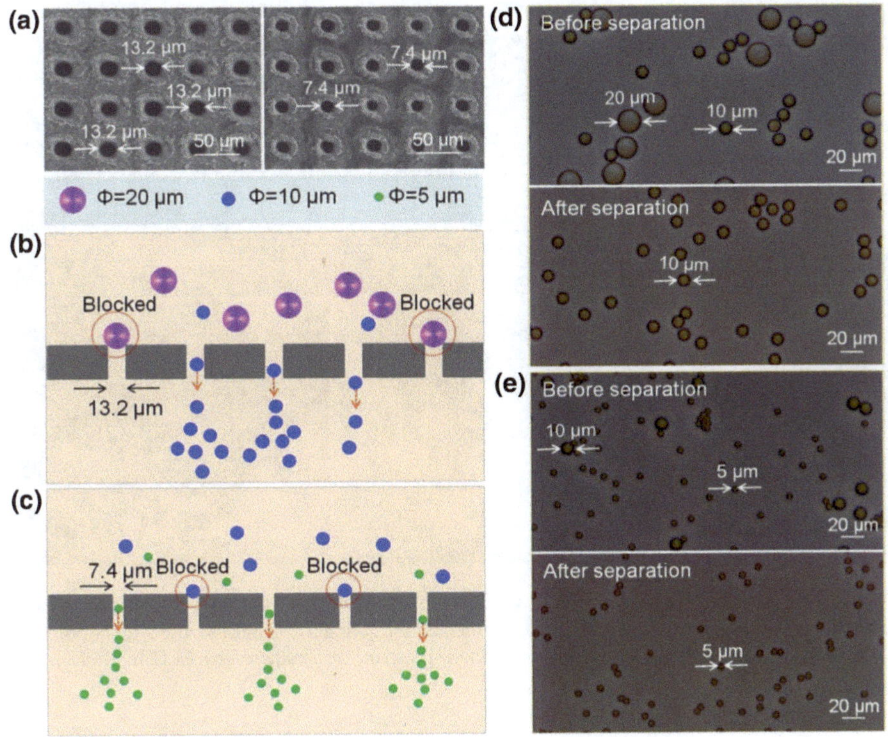

Fig. 8.15 Multifunctional filtration of different diameter SiO_2 particles with micropore-arrayed aluminum foil. **a** The SEM images of the aluminum foils with pore diameters of 13.2 and 7.4 μm, which are used for filtrating particles with sizes of 20, 10, and 5 μm. **b** and **c** The front-view of the schematic diagram for filtrating particles with different sizes. **d** and **e** The optical microscope images of the particle filtration. By comparing the particles before and after filtration, it is well illustrated that the micropore-arrayed aluminum foil has high sorting efficiency

In summary, we have demonstrated a new oil/water separation device based on micropore arrayed ultrathin aluminum foil fabricated by femtosecond laser perforating. This device can continuously separate not only the light oil and water mixtures, but also the heavy oil/water mixtures with high speed and efficiency, as well as recycling the oil resources for reuse. This novel separation strategy overcomes the shortcoming of conventional methods: only separating light oil and water mixtures or heavy oil and water mixtures, cannot continuously separating for retaining oil or water on the separation device, low oil recycling efficiency. The foil displays bright iridescence under the white light irradiation due to the optical diffraction effect of regular micropore arrays. The micropore arrayed aluminum foil also shows robust filtrating performance for particles with different sizes, exhibiting multifunctional applications. The combination of the controllable and flexible femtosecond laser micro/nanofabrication technology with the ultrathin, plastic aluminum foil endow

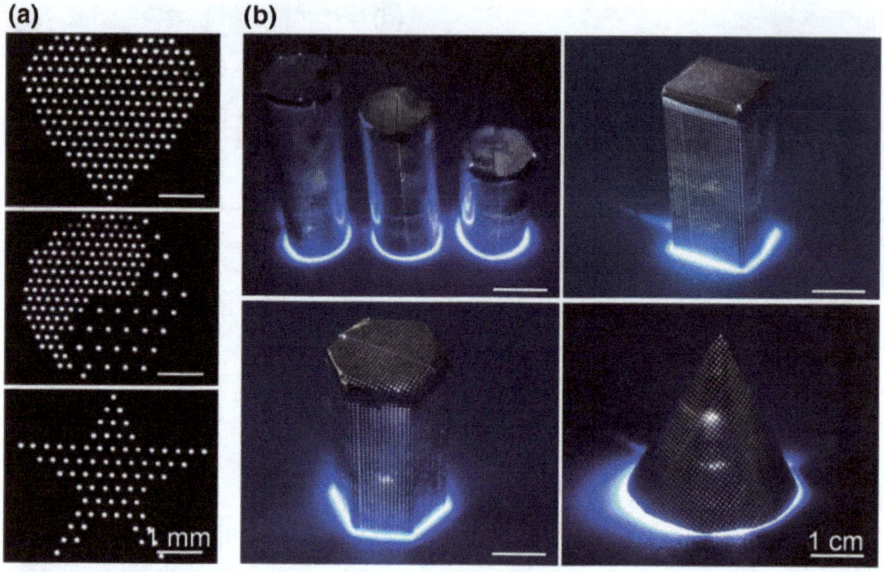

Fig. 8.16 **a** The microscopic photographs of diverse patterns, such as heart, TaiJi, and star-shaped patterns. **b** The assembly of diverse three-dimensional geometric patterns with the as-prepared aluminum foil due to its good plasticity. All these patterns are irradiated by LED lights from the inside

the micropore arrays with particular use in the design of sewage treatment and particle filtration equipment, as well as other multifunctional devices [23].

References

1. Bi H, Xie X, Yin K, et al. Spongy graphene as a highly efficient and recyclable sorbent for oils and organic solvents. Adv Func Mater. 2012;22(21):4421–5.
2. Zhang X, Li Z, Liu K, et al. Bioinspired multifunctional foam with self-cleaning and oil/water separation. Adv Func Mater. 2013;23(22):2881–6.
3. Ge J, Ye YD, Yao HB, et al. Pumping through porous hydrophobic/oleophilic materials: an alternative technology for oil spill remediation. Angew Chem. 2014;126(14):3686–90.
4. Liu N, Cao Y, Lin X, et al. A facile solvent-manipulated mesh for reversible oil/water separation. ACS Appl Mater Interfaces. 2014;6(15):12821–6.
5. Xu Z, Zhao Y, Wang H, et al. A superamphiphobic coating with an ammonia-triggered transition to superhydrophilic and superoleophobic for oil-water separation. Angew Chem. 2015;127(15):4610–3.
6. Shi L, Chen K, Du R, et al. Scalable seashell-based chemical vapor deposition growth of three-dimensional graphene foams for oil-water separation. J Am Chem Soc. 2016;138(20):6360–3.
7. Zhang W, Lu X, Xin Z, et al. A self-cleaning polybenzoxazine/TiO_2 surface with superhydrophobicity and superoleophilicity for oil/water separation. Nanoscale. 2015;7(46):19476–83.
8. Wang B, Liang W, Guo Z, et al. Biomimetic superlyophobic and superlyophilic materials applied for oil/water separation: a new strategy beyond nature. Chem Soc Rev. 2015;44(1):336–61.

9. Yong J, Chen F, Yang Q, et al. Oil-water separation: a gift from the desert. Adv Mater Interfaces. 2016;3(7).
10. Gao X, Zhou J, Du R, et al. Robust superhydrophobic foam: a graphdiyne-based hierarchical architecture for oil/water separation. Adv Mater. 2016;28(1):168–73.
11. Ma Q, Cheng H, Fane AG, et al. Recent development of advanced materials with special wettability for selective oil/water separation. Small. 2016;12(16):2186–202.
12. Liu Y, Ma J, Wu T, et al. Cost-effective reduced graphene oxide-coated polyurethane sponge as a highly efficient and reusable oil-absorbent. ACS Appl Mater Interfaces. 2013;5(20):10018–26.
13. Ruan C, Ai K, Li X, et al. A superhydrophobic sponge with excellent absorbency and flame retardancy. Angew Chem Int Ed. 2014;53(22):5556–60.
14. Wu C, Huang X, Wu X, et al. Mechanically flexible and multifunctional polymer-based graphene foams for elastic conductors and oil/water separators. Adv Mater. 2013;25(39):5658–62.
15. Greenewalt CH, Brandt W, Friel DD. Iridescent colors of hummingbird feathers. JOSA. 1960;50(10):1005–13.
16. Wen Q, Di J, Jiang L, et al. Zeolite-coated mesh film for efficient oil-water separation. Chem Sci. 2013;4(2):591–5.
17. Wang CF, Tzeng FS, Chen HG, et al. Ultraviolet-durable superhydrophobic zinc oxide-coated mesh films for surface and underwater-oil capture and transportation. Langmuir. 2012;28(26):10015–9.
18. Yang J, Zhang Z, Xu X, et al. Superhydrophilic-superoleophobic coatings. J Mater Chem. 2012;22(7):2834–7.
19. Zhang F, Zhang WB, Shi Z, et al. Nanowire-haired inorganic membranes with superhydrophilicity and underwater ultralow adhesive superoleophobicity for high-efficiency oil/water separation. Adv Mater. 2013;25(30):4192–8.
20. Liu N, Chen Y, Lu F, et al. Straightforward oxidation of a copper substrate produces an underwater superoleophobic mesh for oil/water separation. ChemPhysChem. 2013;14(15):3489–94.
21. Gao X, Xu LP, Xue Z, et al. Dual-scaled porous nitrocellulose membranes with underwater superoleophobicity for highly efficient oil/water separation. Adv Mater. 2014;26(11):1771–5.
22. Yong J, Fang Y, Chen F, et al. Femtosecond laser ablated durable superhydrophobic PTFE films with micro-through-holes for oil/water separation: separating oil from water and corrosive solutions. Appl Surf Sci. 2016;389:1148–55.
23. Li G, Fan H, Ren F, et al. Multifunctional ultrathin aluminum foil: oil/water separation and particle filtration[J]. J Mat Chem A. 2016;4(48):18832–18840.

Chapter 9
Conclusions and Outlook

In this work, we first outline the background of bionical technology and its historical developments. Then we take an advanced femtosecond laser micromachining approach to prepare diverse bionic hierarchical structures on materials surfaces, and realize the multifunctions such as bionic structural colors, superhydrophilicity, superhydrophobicity, liquid manipulating, and oil/water separation, and so on.

Based on the research of the relationship between the micro/nanostructures and the laser parameters, a series of structural colors and the diverse displays are studied. By altering the laser pulse overlaps on aluminum surfaces, several single colors which are insensitive to the viewing angle are produced for the formation of nanostructure-covered microgrooves. By skillfully utilizing the correlation of the orientation of periodic ripple and laser polarization, multi-patterns constituted by ripples with different orientations were designed and the diverse display for the desired ones were realized by exquisitely varying the incident light angle and rotating sample angle. Furthermore, we prepared diverse 3D hierarchical structures on metal and silicon surface with liquid-assisted laser micro/nano processing and changed the wettability, thus realizing the controlling of the liquid's motion. In addition, we present regular micropore-arrayed ultrathin aluminum foil fabricated by femtosecond laser perforating. The micropore arrayed aluminum foil shows excellent oil/water separation function and robust filtering performance for particles with different sizes, exhibiting multifunctional applications.

The innovative points of this work can be summarized as: (1) variety bionic functional structures are produced by combining the advanced femtosecond laser micromachining approaches and the frontier bionic science. Many kind special and regular micro/nanostructures can be prepared by one-step femtosecond laser irradiation, due to its high precision, excellent controllability, one-step processing and compatible with various materials. (2) The preparation of composite colorful patterns was developed and their diverse displays were studied, which could provide theoretical and experimental basis for the preparation and practical applications of structural colors. (3) The liquid-assisted laser micro/nano processing method was proposed to induce 3D hierarchical structures. Different structures can be prepared by adjusting the laser parameters and liquid environments. (4) Laser perforating was used for the first time to form controllable micropore possessing multifunction.

G. Li, *Bionic Functional Structures by Femtosecond Laser Micro/nanofabrication Technologies*, Springer Theses, https://doi.org/10.1007/978-981-13-0359-3_9

However, in the scientific research aspect, there are still many important scientific problems that need to be further studied. (1) The exact formation mechanism of femtosecond laser induced micro/nano structures has not been explored although many kinds structures can be accurately controlled by controlling femtosecond laser processing parameters. Hence, it is necessary to study the microcosmic change rules of the structures during the laser interaction with the material surfaces and hereby establish the corresponding thermal dynamic model. (2) The quantitative relationship between micro/nanostructures and the bionic function is not clear. The physical model of the relation between the micro/nanostructures and functions should be established which is based on the analysis of the dynamic influence of the structures on the functions.

In the applied research aspect, in spite of those beneficial properties of the laser induced micro/nanostructures, there are still several limitations in implementing these materials in engineering settings. First, all the experiment and test are conducted in the laboratory, which are not suitable for harsh environments. Real applications require excellent chemical stability and mechanical robustness. Moreover, the current preparing method is still inefficient and expensive, which can't meet the needs of practical applications. Thus, cheaper, accessible materials and fabrication methods should be sought to translate the lab demonstrations to the market.

Considering the compelling advantages inherent in femtosecond laser micro/nanofabrication and the dramatic development in the correlative science and engineering community, we are confident that preparation of bionic functional structures by femtosecond laser micro/nanofabrication technologies will embrace a bright future.

Kolibri 360 GmbH
Ferdinand-Jühlke-Straße 7
99095 Erfurt, Deutschland
produktsicherheit@kolibri360.de